Modern Techniques in Fish Handling and Processing

Modern Techniques in Fish Handling and Processing

K.D. Bhardwaj

CYBER TECH PUBLICATIONS
4264/3, Ansari Road, Daryaganj, New Delhi-110002 (India)
Ph.: 011-23244078, 011-43559448 Fax: 011-23280028
E-Mail: cyberpublicationsdelhi@yahoo.com
Website: www.cybertechpublications.com

MODERN TECHNIQUES IN FISH HANDLING AND PROCESSING

K.D. Bhardwaj

First Edition 2011

Published by :

G.S. Rawat for **Cyber Tech Publications**
4264/3, Ansari Road, Daryaganj, New Delhi-110002 (India)
Ph.: 011-23244078, 011-43559448 Fax: 011-23280028
E-Mail: cyberpublicationsdelhi@yahoo.com
Website: www.cybertechpublications.com

CONTENTS

CONTENTS

Preface

According to the recent FAO Book on by-catch and discards, the weighted discard rate of fisheries is estimated at 8% and the yearly average discards are estimated to be 7.3 million tons that are directly thrown to the seas. They are constituted by non targeted species and by residues due to fish transformation on-board. After landing, this wide variety of marine species of animals is converted into seafood. Of those animals, only a portion is usually separated from the carcass and used as food. The remainder is by-product, often rich in protein, which can (should!) be processed into useful products. By-products from fisheries, included fish farming consist of viscera (liver, roe, stomachs, etc.), heads, backbones, trimmings and rejected fish from processing. These by-products are generated when the fish is headed, gutted and further processed. The raw materials that come from traditional fisheries and aquaculture can be regarded as a great and valuable source of proteins both for animal and human nutrition. Today most of the by-products are used as raw materials for feed production; such as fish meal, fish sauce, fish silage and feed for animals. However, as presented in this book, promising opportunities exist for upgrading marine by-products and underutilized fish by using mild processing techniques to convert them into protein and peptide ingredients both to improve yield in traditional seafood and to be used as nutraceuticals or in functional foods. Many mild techniques like pH shift, fermentation, enzymatic hydrolysis, filtration, centrifugation and spray and freeze drying can be used in the processing and production of valuable products. Indeed, it is estimated that if we succeed to utilize more of the by-products as food for humans and as ingredients in foodstuff, health foods, nutraceuticals, pharmacy, cosmetics etc., the value adding may increase by 5 folds. However, adequate

handling and pre-treatment practices for discards and wastes both on board and on shore (auctions) including separation, classification, stabilization and conservation are required in order to preserve valuable components and to allow further added value. Some examples of processes (enzymatic hydrolysis, membrane filtration, centrifugation,...), products (proteins, peptides, lipids, aroma,...) and applications for those resulting products (feed, food, functional food, nutraceutical, health,...) are given into this book but many others exist. Indeed, it is reasonable to expect that innovation will continue and will help industry to turn waste into by-products and by-products into products. Can we imagine that considering wastes as resources will solve the problem of wastes and will preserve the resources? A unique feature of the current book is that it brings together, in one place, information of the entire field of marine by-products soft-processing and utilization. We have incorporated new and updated information throughout this book. This book will be useful for anyone interested in marine by-product soft-processes or marine by-product utilization, individual interested in recycling and those interested in a sustainable example of reutilization of valuable material. Careful consideration was given to selecting the contributors of this book. The authors were all belonging to the network SEApro (Sustainable Exploitation of Aquatic PROducts, www.seapro.fr) dedicated to the fisheries wastes upgrading for promoting a total utilization of the fish in order to be in accordance with international trends in development of sustainable fisheries. Authors are all recognized as experts in their respective areas. I would like to thank all the contributors for their hard work and tolerance for my sometimes severe demands. Experts have refereed all papers and I wish to express my gratitude for their referee work.

— *Author*

Introduction

This note summarizes the main ways in which fish is handled, processed and distributed in Britain, both at sea and on shore. It is principally written for those concerned in industrial training; more detailed information on particular aspects of fish handling and processing is given in other notes in this series, and the reader is referred to some of these where appropriate. The principles and practice of handling and processing are dealt with at length in the book *Fish Handling and Processing,* Second Edition edited by G. H. O. Burgess and others, and published by HMSO 1965, price £2.

The topics outlined in this note include the handling of wet fish, smoked fish, frozen fish, dried and salted fish, canned fish and shellfish, and the manufacture of fish meal and other byproducts.

HANDLING WET FISH

Much of the fish landed in Britain is preserved by chilling in ice from the time it is caught until it reaches the consumer. Fish preserved in this way is known as wet fish. A growing proportion of the catch is frozen at sea immediately after capture, and considerable amounts of iced fish are frozen at the ports after landing; these operations are described under freezing and cold storage.

HANDLING WET FISH AT SEA

The length of voyage of fishing vessels storing the catch in ice may range from a few hours for small inshore vessels to about three weeks for the largest distant water trawlers. White fish, that is those species in which most of the fat is in the liver and the flesh is lean, are handled in much the same way on all sizes of vessel. The catch is released from the net on to the deck, gutted immediately, washed, and stowed with ice in boxes or compartments below deck.

Gutting of round fish like cod, haddock and whiting means slitting the belly from throat to vent, removing the liver and cutting out the guts to leave the belly cavity empty. This operation is traditionally done by hand with a knife, but gutting machines are coming into use on both large and small ships to make the task of the fisherman easier. Gutting helps to preserve the fish by removing the main source of spoilage bacteria and digestive juices which attack the flesh of the fish after death. On the larger fishing vessels the livers are cooked in steam boilers to extract the liver oil, but on small boats the livers are discarded with other offal.

The gutted fish are washed to remove traces of blood and debris, and to wash away most of the bacteria present on the skin and in the gills of the fish. The washing equipment on small boats may be simply a hose and an open mesh basket, but on large trawlers a more sophisticated washing tank with circulating water is in general use. In these washers the fish are discharged over a weir and down a chute to the fishroom below deck.

Fishrooms for iced fish are mainly of two types, either an undivided hold in which the catch is stowed in boxes, or a hold divided by partitions into a number of sections called pounds in which the catch is stowed on portable shelves. The principle of stowage is the same; the fish are in shallow layers completely surrounded by ice, whether on a shelf or in a box, so that they are cooled rapidly to ice temperature and kept close to 0°C throughout the voyage. About one part of ice to three parts of fish by weight is required to protect fish for up to 5 days; one part of ice to two of fish is needed for longer voyages. White fish, promptly gutted, washed, and stowed in ample ice, will keep in first class condition for 5-6 days, become stale after 10-12 days, and are unlikely to be edible after 15-16 days.

Boxed stowage is usual on smaller fishing vessels, and the practice of boxing is gradually being extended to larger ships, since the method has a number of advantages including delivery of the fish to the merchant undisturbed by rehandling at the port market, ease of identification of size, species and time in ice, and avoidance of damage and loss of weight during stowage.

Ice plants, particularly older ones at the larger ports, supply crushed block ice to fishing vessels, but more recently built plants, particularly at smaller ports where ice was not locally available in the past, usually deliver small, smooth pieces of ice known as flake ice. Flake ice is normally bulkier than crushed block ice, but weight for weight the cooling capacity of all types of ice, made by any method and from hard or soft fresh water, is the same.

The fish are cooled when heat is absorbed by the surrounding ice, which is thus melted. Further cooling is obtained when the cold meltwater trickles down between the fish. The fishroom temperature is ideally kept slightly above 0°C in order to allow the ice to melt slowly, but is not kept so high as to waste the ice. To this end, most fishrooms on large vessels are completely insulated, and fishrooms on small boats often have partial insulation. Mechanical refrigeration plants are installed on one or two inshore vessels and on some, but by no means all, larger ones; their main purpose is to conserve ice on the outward voyage, and to keep the fishroom air cool during fishing; they have little or no direct effect on the stowed fish, which depend for cooling almost entirely on the surrounding ice.

Fatty fish, like herring, sprats, mackerel and pilchards, i.e. those containing a good deal of fat or oil, are not normally gutted at sea because their small size and the large numbers in which they are caught make this impracticable in the time available on typically short voyages to grounds not far from the port of landing. They are usually put below straight from the net, and iced in boxes. The keeping time of fatty fish in ice is much less than for white fish; the attack by bacteria and digestive juices is much more rapid because the fish are ungutted, and the fat absorbs oxygen to produce rancid flavours and odours. Herring for example are normally required to be in the hands of the port processors within 1-2 days after catching to give a first class product, although for some outlets it is possible to keep herring with a low fat content for 4-5 days in ice. Stowage at sea in refrigerated sea water is a possible alternative to ice as a means of rapidly cooling large quantities of small fatty fish, although the method is not yet in general use in the UK.

The following Advisory Notes, expand the information on handling wet fish at sea: 4 *Take care of your catch;* 7 *The protection of wood in fishrooms,* by J. J. Waterman; 11 *Handling inshore fish,* by J. J. Waterman; 15 *Bulking, shelfing or boxing?* by J. J. Waterman; 21 *Which kind of ice is best?* by J. J. Waterman; 32 *Superchilling,* by J. J. Waterman and D. H. Taylor; 33 *The cod,* by J. J. Waterman; 42 *Fish for caterers and friers,* by J. C. Early and R. Malton; 44 *Handling fish before canning,* by R. McLay; 47 *Handling and processing saithe,* by J. G. M. Smith and R. Hardy.

HANDLING WET FISH ON SHORE

The catches of British fishing vessels are normally auctioned to fish processing firms at the fishing ports. Most of the large ports are on the east coast, including the three biggest. Hull, Grimsby and Aberdeen. Covered quayside markets are provided at almost all the ports, and merchants' premises are usually adjacent to, or a short distance from, the markets. At a few remote landing places, particularly on the west coast of Scotland, where no local processing facilities are available, the catches are discharged and consigned by road to the processing centres on the east coast.

Comparatively small amounts of white fish are dispatched unprocessed, repacked in fresh ice, to inland destinations but most white fish is filleted in premises at, the port. Although machines are available for filleting most species, a large proportion of the catch, particularly when handled by the smaller processing firms, is still filleted by hand. The fillets, which constitute roughly 40-50 per cent by weight of the gutted whole fish, are either packed in ice in non-returnable boxes and sent to inland wholesalers and retailers, or are further processed at the port, mainly by smoking or freezing, or both.

A high proportion of the herring catch goes to the kipper trade, and herring for this outlet is split or filleted by machine for making kippers and kipper fillets respectively.

The traditional container for inland carriage of fillets has been the non-returnable wooden box, mainly in units of 7 and 14 lb, but this has been superseded in many large firms by a waterproof fibreboard box and, to a lesser extent, by an expanded polystyrene box, which has some advantage as an insulated container but is more susceptible to damage due to rough handling during transit. Fillets should be kept close to 0°C during transit, and in good practice the fillets are packed in thin layers with ice top and bottom and a little more ice sprinkled among the fish.

Fish is nowadays mostly carried by road transport from the ports and the biggest companies have their own fleets of insulated and refrigerated vehicles carrying fish either to inland distribution depots or direct to customers, while many smaller merchants at the ports share a long-distance transport pool.

Although the better carriers use vehicles with adequate insulation, often supplemented by mechanical cooling units, some fish is still carried under less satisfactory conditions on open lorries; much greater reliance has then to be placed on ice and on the insulating properties of the boxes to protect the contents during distribution. Under the best conditions, wet fish can be on sale to the consumer anywhere in Britain within twenty four hours of landing; a typical timetable would be discharge from the fishing vessel in the early hours of the morning, sale to the port merchant and a short journey to his premises at about 8 am, filleting, packing and icing by midday and dispatch by road from the port in the afternoon to arrive at an inland depot in the early hours of the next day, from where it is delivered to the retail shops in time for that day's trading.

Most retailers keep the bulk of their supplies in chillrooms on the premises and only display a selection on ice, often in refrigerated cabinets.

The following Advisory Notes expand the information on handling wet fish on shore: 1 *The care of the fishmongers fish,* by G. H. O. Burgess; 3 *The handling of wet fish during distribution;,* 10 *Fishworking premises- materials and design,* by J. J. Waterman; 12 *Fish display in retail shops;* 16 *Non-returnable fish boxes,* by J. Wignall; 17 *Measures, stowage rates and yields of fishery products,* by J. J. Waterman; 23 *Control of flies in fishmongers' shops;* 42 *Fish for caterers and friers,* by J. C. Early and R. Malton; 45 *Cleaning in the fish industry,* by I. N. Tatterson and M. L. Windsor.

Smoked Fish

Fish is smoked nowadays mainly to give it a pleasant flavour rather than to preserve it. Present day products are therefore only lightly salted and smoked and will not remain edible for much more than a week at ordinary temperatures. The smoking process consists of passing wood smoke over the surface of the fish, in a kiln. Most British products are cold smoked, that is the fish remains uncooked and the kiln temperature does not rise above 30°C. Typical smoked products are the finnan haddock, smoked cod fillet, the golden cutlet and the kipper. Fish that are hot smoked are cooked during the process; the kiln temperature may be as high as 80°C and the fish temperature may reach 60°C. Some hot smoked products in this country are sprats, eels, trout, buckling made from herring, and Arbroath smokies made from small haddocks.

Two types of smoking kiln are in general use, the traditional chimney kiln and the Torry mechanical kiln. It is estimated that more than half of the smoked fish made in Britain is now produced in mechanical kilns, and the proportion is continually increasing.

Before smoking, the fish are immersed in a brine solution. This assists in removing some of the water in the fish, thus tending to firm the flesh. The salt imparts a flavour to the product, but concentration and purity of the salt are extremely important and require to be carefully controlled. A 70 to 80 per cent saturated brine is used in most modern smoke cures.

Following the salting treatment, pre-drying of the fish is required in order to remove some of the moisture prior to smoking. For this purpose, the fish are hung to drip on open racks.

The source of smoke is almost universally a smouldering fire of hardwood chips and sawdust; although more sophisticated smoke producers have been made from time to time, and are used for smoking other foods, these have so far made little impact on the fish trade. In the traditional chimney kiln, the open fires are at the base of a tall, brick-built structure in which the fish are hung on rails of various types called banjoes, speats or tenters, and thus exposed to the rising smoke and warm air. The repositioning of the fish during smoking and eventual removal of the finished products are slow hand operations which require the services of a skilled craftsman in order to produce a satisfactory article.

In the mechanical kiln, the fires are contained in separate fireboxes, and the smoke is blown horizontally through trolleys holding fish in the kiln; the fish may be hung on rails or laid on trays, either of which are supported in the trolley. The temperature and speed of the mixture of smoke and air is carefully controlled to give a uniform product throughout the kiln in a much shorter time than is possible in the chimney kiln. Fish handling is much reduced using the mechanical kiln which can also readily be incorporated in the factory production line. Partial drying as well as smoke deposition is an essential part of the smoke curing process; typically a kipper which should lose about 14 per cent in weight during smoking will require 6-12 hours in a traditional kiln, but only 4 hours in a mechanical one.

Most cold smoked fish products are only lightly coloured by the smoke, and so a permitted dye is normally added to the brine bath through which the fish pass before going into the kiln, in order to enhance the appearance of the finished product.

FREEZING AND COLD STORAGE OF FISH

It is perfectly feasible to keep fresh fish for many months without perceptible change in the eating quality by rapidly freezing it soon after catching and then storing it at a suitably low and constant temperature. With this method of preservation, the thawed product is virtually indistinguishable from the best fresh fish. Freezing has revolutionized the fish processing industry in Britain since the Second World War.

FREEZING FISH

Heat is removed from the fish in the freezing process either by surrounding the fish with a stream of cold air, by placing the fish in contact with a cold surface or by spraying with certain liquid refrigerants. Three main types of freezing plant are used that employ these techniques, the air blast freezer, the plate freezer and the immersion freezer.

The air blast freezer is essentially a tunnel in which a fast-moving stream of very cold air is blown over the fish, which are placed on trolleys or on a moving belt. The air is usually at a temperature of-30 to -40°C and moving at about 5 m/s. The air blast freezer is most suitable for a wide range of sizes of fish, and for products of irregular shape.

The plate freezer is more compact than the air blast freezer, and is most useful for handling fish products that are uniform in thickness and that have a reasonably flat surface which can make good contact with the cold plates. Two versions of the plate freezer are in general commercial use, the horizontal and the vertical types. The horizontal plate freezer, used mainly in land installations, handles many of the catering and retail fish products that are already packed in cartons prior to freezing. Trays of packs are slid between pairs of horizontal plates, the plates are closed tightly on to the packs by hydraulic pressure to make good contact, and a cold refrigerant is circulated through serpentine passages within the plates; a retail pack 3 cm thick takes about an hour to freeze.

The vertical plate freezer, which was originally designed in the 1950s for use on fishing vessels, is employed for freezing large blocks of whole fish. The fish are packed between pairs of plates, usually without any wrappings, and the plates moved slightly towards each other to compact the block and ensure good contact. Liquid refrigerant is circulated through the serpentine passages within the plates until the fish are frozen; the complete process for a block of whole cod 10 cm thick takes about 4 hours including loading and unloading time, with refrigerant at -40°C.

The immersion freezer is not used as much as the plate and blast freezers in the British fish industry; the main difficulty is the limited choice of liquids that are both good refrigerants and suitable for use in direct contact with foods. Brines are sometimes used, but the fish may take up too much salt. Liquid nitrogen is used successfully in one type of immersion freezer, which subjects the product to a spray of liquid nitrogen as it passes through a tunnel.

The freezing process cannot improve the quality of fish; therefore the best frozen products are those made from first class raw material. This applies particularly to whole fish which, after thawing, are likely to be subjected to further processing. Whole iced cod, for example, when frozen not later than 3 days after catching can on thawing be treated in the same way as very fresh fish, but cod that has been delayed longer than this, or has been kept uniced, is unlikely after freezing and thawing to yield fillets of a high quality. Whilst some species, flatfish for example, can be kept in ice a little longer than cod before freezing without impairing the quality of the thawed product, others like haddock and hake do not keep so well. Fillets of most white fish species can be taken from whole fish 5-6 days in ice and frozen to give a high quality thawed product.

The freezing process should always be completed as rapidly as possible, not only to increase output of the equipment, but also to reduce the time in which bacteria and digestive juices are still able to attack the fish; bacterial action ceases below about -10°C, and the activity of enzymes is reduced as the temperature falls. There are marked changes in texture and flavour when fish is frozen very slowly at temperatures only a little below 0°C. The final temperature of fish being frozen should be that at which it is to be stored, namely -30°C; this ensures that the frozen product imposes no extra heat load on the cold store since this is designed only to keep the product cold and not to freeze it.

Freezing at Sea

On freezer trawlers handling whole fish, the catch is brought in over the stern, discharged to an enclosed processing deck, gutted by hand or machine, washed and bled in cold sea water and conveyed forward to the freezers, where blocks of about 50 kg are produced in vertical plate freezers and transferred to cold storage rooms running at -30°C. These ships make voyages of 4-8 weeks, carry a crew only two or three more in number than their counterparts handling iced fish, and bring back about 500 tonnes of frozen cargo for transfer to shore cold stores at the ports.

Factory trawlers that fillet the catch before freezing are less numerous than trawlers which freeze whole fish. On the former the gutted fish are filleted mainly by machine, then the fillets are packed into blocks of about 10-15 kg and frozen in horizontal plate freezers before being transferred to the ship's cold store. Compared to the situation with a trawler freezing whole fish, the crew has to be larger to cope with factory operation at sea, and the quality of fillets frozen at sea as opposed to whole fish frozen at sea depends much more on careful handling and processing, but on the other hand the storage capacity of the ship in terms of frozen edible portion is considerably increased.

It is likely that the distant water fishing fleet will gradually change over completely from the stowage of wet fish in ice to the freezing of fish at sea.

Cold Storage of Fish

Temperature of storage is the most important single factor affecting the storage life of frozen fish. Almost all cold stores for fish that have been built in recent years are designed to operate at -30°C, at which temperature the products will keep in first class condition for several months. Lean fish, such as cod and haddock, when stored for long periods at too high a temperature or in fluctuating temperatures can have marked changes in texture and flavour when thawed out. The thawed flesh may feel rubbery and appear dense white instead of translucent. After cooking they are found to be tough and fibrous or stringy.

Other causes of change in cold stored fish are dehydration and oxidation. Dehydration is kept to a minimum either by glazing unwrapped frozen fish before storing them, that is covering the surface with a skin of ice by dipping them quickly in cold water, or by packaging the fish in a material that is a good barrier against the passage of water vapour, for example polyethylene film. Fatty fish like herring are particularly prone to absorb oxygen from the air and so become rancid; these are therefore wrapped in a material that forms a good oxygen barrier. In addition the space between the package and the contents may be evacuated to reduce even further the risk of rancidity. The ideal packaging material for fish products is often a laminated film combining the desired properties of two or more plastics. The cold storage chain is maintained in distribution by the use of insulated, refrigerated vehicles or containers, both of which operate at about -20°C, and frozen food cabinets, again at -20°C, in shops and catering premises.

Thawing Frozen Fish

The growth of quick freezing in the fish industry, and particularly the production of large blocks of whole fish for subsequent processing, has made necessary the development of thawing plant. There are two main types of equipment, those in which the fish are heated in a warm air stream or in warm water, and those which directly use heat generated by electricity. The method most used for large blocks of sea frozen fish is air blast thawing; the fish are conveyed through a stream of moist moving air at about 20°C until they are thawed enough to permit filleting or other processing. The fish temperature should never exceed 20°C.

DRIED AND SALTED FISH

Bacteria and moulds generally cannot grow in the absence of water and hence drying can be used as a means of preservation. Salt, if present in sufficient strength, will slow down or prevent bacterial spoilage of fish. Drying or salting, or a combination of both, have been used in the fish industry for centuries but nowadays only a very small proportion of the catch in Britain is processed by these methods.

A few companies still make dried salted fish from cod and related species, by heading and splitting the fish, removing most of the backbone, and stacking the fish in piles with layers of salt between them. The juices withdrawn from the fish by the salt are allowed to run away and, after frequent restocking over a period of months, the water content is further reduced by drying the fish in a heated chamber until the moisture content is somewhere between 10 and 30 per cent.

Some herring are pickle cured, particularly in north-east Scotland and the Shetlands, although this export trade is a mere shadow of what it was earlier this century. The whole herring, having been lightly sprinkled with salt while awaiting processing, are first gibbed by hand or machine, that is the gills, long gut and stomach are removed. They are then packed in barrels with a layer of salt on each layer of herring until the barrels are full. After a day or two, when the herring have shrunk appreciably, the barrels are topped up with further layers of herring and salt, the lids are fitted and the barrels then laid on their sides for 8-10 days. After this period the barrels are up-ended, the lids removed, and the blood pickle from the upper half drained off through the bunghole. To make ready for storage the barrels are finally topped up with fish, the lids replaced and blood pickle poured through the bung-holes until all spaces are filled. Klondyking is the name given to another method of preserving ungutted herring using salt. The name is thought to originate from a method developed about the same time as the famous Gold Rush of 1897.

The herring, contained in baskets, are sprinkled with coarse salt, approximately 175 kg of fish treated with about 10 kg of salt, as they are tipped into a wooden box. Ice is then put on this mixture offish and salt, about 100-125 kg of ice to every 175 kg of mixture. Herring treated in this way could be kept in edible condition for approximately 1 week.

CANNED FISH

The canned fish industry in Britain is a small one, and the range of products is confined mainly to herring, sprats and pilchards packed in either tomato sauce or vegetable oil.

The process for herring in tomato sauce is typical. First the fish are nobbed by machine, that is the head and gut are removed. They are then immersed in saturated brine for up to 30 minutes and packed by hand into oval cans holding 200 g of fish. The tomato sauce is added, the can lids are lightly clipped on and the cans exhausted in steam for 10-15 minutes. The object of exhausting is to produce a partial vacuum in the headspace of the can which is not filled either with solids or liquids. The cans are then sealed, washed, and heat processed in steam at 115ºC for 55 minutes. After cooling, the cans are stored for about a month, labelled and packed in outer cartons for dispatch. The heat processing stage is critical, and is designed to inactivate all bacteria and enzymes present and in particular to destroy any harmful organisms.

There are a number of imported canned fish products that could be manufactured in this country from British-caught fish, and the canning industry is examining the possibilities of expanding their range.

HANDLING SHELLFISH

The shellfish industry, although only a small part of the fish industry as a whole, has grown considerably in recent years, and the products are generally high value ones. The principal species in order of importance are Norway lobster or scampi, lobster, scallop, crab, cockle, crawfish and oyster. Mussels and shrimp make only small contributions to British landings at the present time, but their fisheries are capable of considerable expansion.

The Norway lobster is landed either whole or headless in ice, and is often frozen in the shell while awaiting processing. The meats are extracted from the thawed tails either by hand peeling or by blowing out with air or water jets. The peeled meats are frozen individually, either in an air blast freezer or a liquid nitrogen freezer, glazed and bagged and put into cold storage.

Lobsters are still distributed live inland; the few that are processed are normally cooked and then frozen whole. Crabs, which do not travel well, are processed close to the points of landing. They are boiled whole, and the meats are then extracted from body and claws by hand. The white and brown meats are frozen separately and then wrapped and cold stored. A small amount of crab meat is also canned in Britain, and small amounts of cooked whole crabs are distributed chilled to retailers.

The oyster is marketed live in shell, and the crawfish catch is mainly exported live to the Continent. The scallop, and its near relative the queen, are frozen in shell on arrival at the port processing plant, thawed as required and the meats removed by hand with a knife. The meats are frozen individually, usually in an air blast freezer, and bagged for cold storage; they are mainly exported to the United States. Cockles and mussels are boiled in the shell at the ports; the extracted meats may be distributed either chilled, frozen, or packed in jars in brine or vinegar.

Shrimps are cooked and peeled soon after catching, and the meats distributed either chilled, frozen or potted in butter; alternatively the whole raw shrimp are iced and then frozen for subsequent processing.

FISH BYPRODUCTS

When the edible flesh is removed by filleting from whole fish landed for human consumption, the remaining heads, skeletons and other processing waste are disposed of by making fish meal. In addition whole fish, both white and fatty, that are surplus to market requirements or landed specifically for the byproducts industry are also converted into fish meal and oil. White fish and white fish offal has most of the water removed by cooking and drying, and the dried material ground and bagged for use as a high protein animal food, particularly in the pig and poultry industries. The yield is roughly one tonne of meal from 5 tonnes of raw material. Fatty raw material is pressed after the cooking stage and the press liquor, a mixture of oil, water and some solids, is further processed to separate the oils and solids. The oil-free solids are dried and ground along with the bulk of the solid material from the press. The refined fish oils make a valuable contribution to the manufacture of edible products like margarine.

Fish meal made from suitable raw material under hygienic conditions can be eaten by humans, and its possible use as a fish concentrate protein added to the diet of protein-deficient peoples in the developing countries is being investigated in many parts of the world.

Waste from the British shellfish industry is occasionally accepted in small quantities by the fish meal manufacturers, but most of the shells are at present unsuitable for the manufacture of byproducts.

Only a small part of this country's requirements for fish meal are met by the British industry; the possibility of reducing imports by increasing the British catch of species unwanted for human consumption and converting it into fish meal is at present being explored.

2

Planning and Engineering Data

INTRODUCTION

This document presents selected planning and engineering data on fish boxes and containers used in fish handling and marketing. The purpose of this document is to provide the industry, especially in developing countries, with a practical, up-to-date set of easy reference data of a technical and economic nature, particularly data required by those concerned with planning and costing of fisheries industrial development programmes and those assessing the technical feasibility of relevant projects.

The suitability for use by governmental offices responsible for various aspects of commercial fisheries, specialized schools, training courses, etc., is also taken into consideration.

This document does not pretend to include all the relevant data on fish boxes and containers. The limited funds made available for this study has made it necessary to cover parts of the material less deeply than would be desirable.

Some information such as price of materials is subject to rapid change so this report aims more at describing general concepts. More specific information in these areas can be sought by people using the report.

The publication is not the result of original research, but is based on information available from manufacturers of equipment, from previously published material by numerous organizations, particularly FAO of the UN, Norwegian Standard Organization, Directorate of Fisheries, Norway and some foreign institutions.

CONTAINERS IN THE FISHING INDUSTRY

Fish as Raw Material

A large proportion of the world population is dependent on fish as a source of animal protein. The flesh of fish contains many nutrients which enable consumers to avoid malnutrition.

Table 1 The nutrients in fish per 100 g edible material

Type of fish	Water(g)	Protein(g)	Carbohydrate(g)	Fat(g)	Energy(g)
Lean fish	78	18	-	0.3	7.5
Fat fish	58	20	3	14	205

It should be realized however that fish is a hunted resource and is subject to physical damage during catching and handling, biochemical deterioration due to enzyme action, and bacterial spoilage. The three key factors in maintenance of fish quality are:

- Cleanliness
- Careful handling
- Fast cooling (effective preservation)

Control of these three factors will enable the industry to avoid wastage of time, effort and money and, in the wider context, will ensure that maximum use is made of this natural resource to offer fresh food of high quality to the public.

DETERIORATION OF FISH QUALITY

As previously noted. fresh fish is a highly perishable raw material that will deteriorate rapidly if it is poorly handled. Rough treatment such as trampling underfoot, thrawing, or crushing during stewage will cause rapid visible deterioration.

This will result in bruising and squashing of the flesh and splitting of the guts which contaminates the meat with the gut contents including digestive enzymes that attack the muscle structure and bacteria.

The following table shows the effect of temperature of shelf life on fish. It shows the time taken for freshly caught fish to become unfit for any use when stored at various temperatures.

Table 2 Shelf life and temperature

Temperature	Shelf Life
-1°C	20 days
0°C	15 days
+6°C	6 days
+12°C	3 days
+18°C	11/2 days
+24°C	1/2 day

The effect of temperature can be exacerbated by contamination such as exposure to dirty water and ice, dirty containers and surfaces in fish holds, wharves or factory floors and tables.

In general fish is one of the safest of proteinaceous foods and only causes food poisoning when it becomes contaminated by contact with sail vermin and man. Effective measures should be taken to avoid such contamination.

THE IMPORTANCE OF CONTAINERS IN FISH HANDLING

Containers of various types, sizes and of different materials are used all over the world to hold fish both on board vessels, in processing, during transport and under general storage. Poor handling and lack of suitable containers leads to as much as 20-30% spoilage in many countries.

Containers are used to perform the following functions:

1. ease the handling of small and large quantities of fish;
2. simplify and increase the speed of unloading/loading and transportation of raw material;
3. protect the fish against physical damage contamination and other deteriorating factors;
4. offer a suitable packing unit for fish and ice;
5. contain the fish under such conditions that it reaches the buyer in the best possible condition;
6. help to protect the raw material against natural deteriorating effects;

7. help to make maximum utilization of resources and to achieve optimum economical results through the whole system of handling from harvest to consumption.

In the whole system of correct fish handling the container is only one factor in the process. Knowledge of handling, necessary regulations/laws and the specific behaviour of fish meat under various conditions are also important points to consider.

Various packing materials and containers have their limitations. It is therefore important to choose the right container system at the right time and to use it in accordance with laid down objectives.

In conclusion there is a great variation in the size, design and construction of containers used in various parts of the world. The materials used depend on the type of fishing, size of vessel, type of road transport, degree of organization of the industry, value of the catch, and sometimes local traditions.

Definition

In this report it is felt necessary to define a difference between fish baskets, boxes and containers.

Fish Baskets

Originally a container of plaited or woven material. Most often made of cane or some other strong plant fibre. This type is still commonly in use in most developing countries. In the industrialized fishing nations the traditional plant fibre has been replaced by plastic. It has been common to make the fish basket in sizes ranging from units which could take around 10 kg up to those able to take 100 kg.

The basket has gradually been losing ground to the fish box due to its unsuitable shape for storage (in fish holds, cold stores, etc.) and for mechanized handling, by fork lifts and similar systems.

Fish Boxes

Generally these are uninstalled containers with a capacity from 10 litres to around 100 litres. Each unit is designed to enable handling by 1 or 2 persons.

The fish box has its application on board fishing vessels, in warehousing, processing, transport and at the fish market.

The 3 main areas for use of boxes are as follows:

a. Tote boxes used aboard fishing vessels and for transport of fresh fish and storage of raw material awaiting processing. Boxes for these uses have to be strong, stable, easy to stow, simple to clean and preferably of the returnable type.

b. Boxes for internal use in the processing industry. Normally such boxes are of a lighter and smaller type. Since these units will be containing semi-processed or processed products the hygienic requirements will be high. Most of them are of the returnable type and made of aluminium or plastic. Size, 10 to 30 litres.

c. Boxes or packing material for marketing. At this stage non-returnable boxes are generally used. The main requirement is that the packing material should have sufficient strength to protect the fish until it reaches the consumer, but be low priced. Most often boxes for this use are made of styrofoam, plastic, light wood, or fibre board.

Fish Containers

These are insulated or uninstalled units often tailored to a special purpose or use. Most often they have their application in storage on board vessels and ashore and in transport of fish.

a. Containers used for cooling with ice. Preferably such types should be insulated and have an insulated lid. Size may vary from 50 up to 1000 litres. In this type of container fish can be held for a considerable period of time if sufficient ice is used. They have their application on board fishing vessels, in transport of fish (sea, land, air), as storage units for the processing industry and in the fish market.

b. Self contained cooling containers. Over the past 20 years containers used for transport of various products have been standardized. For transporting products that must be held at low temperatures, insulated, refrigerated containers have been developed. These containers have been designed to standards set by the International Standarization Organization (ISO) so as to fit into the general goods transport system developed for sea, road, railway and air transport.

Containers used for transport of fish can take from 5 to 30 t and are mostly used for long distance transport.

The general requirements for all packaging material used for packing of food, fish included, are that the packaging material shall be clean, undamaged and suitable for the purpose.

The physical strength of the material must be so good that the commodity is protected against pressure and shocks in the period the commodity is stored.

Further from a chemical point of view the materials must fulfill the following requirements:

- must not contain or produce poisonous or health damaging substances or compounds that can be transferred to people or animals eating the fish;
- must not produce smell, taste or colour which can contaminate the product.

To fulfill these requirements it is important that a set of official rules be formulated and that these rules are enforced by the appropriate authorities.

MATERIALS TO BE USED IN FISH BOXES AND CONTAINERS - PROPERTIES AND SUITABILITY

In manufacturing of fish boxes and containers, a great number of different materials have been used. The main categories of materials are metals or alloys, natural materials like wood, bamboo, fibre board, etc., and synthetic materials (plastics).

The most common materials used in manufacturing fish boxes and insulated fish containers are wood, aluminium alloys, HD-polyethylene (high-density polyethylene), expanded polystyrene ("styropor") and solid fibre board. Even if other materials such as steel, bamboo, etc., are in common use in many developing countries, the properties of these materials are not very suitable. Steel is very corrosive and is three times denser than aluminium.

Bamboo, as it is normally used, is a material which presents hygiene problems. This description of materials therefore mostly covers the common ones mentioned above.

Materials are rated on their cost, ease of constructing suitable containers and physical and chemical proportion.

PROPERTIES OF MATERIALS USED FOR FISH BOXES AND INSULATED CONTAINERS

Plastic

Fish boxes and containers made from HD-polyethylene are superior to those made from other materials, but they are rather expensive. The fabrication is inexpensive, but the material carries a relatively high cost (depending on oil prices). Production must be carried out in large numbers to get a price which competes with other materials. The expected life is 5-7 years when properly handled. Life is longer for boxes designed to facilitate stocking.

The material is impervious to liquids which facilitates cleaning and reduces opportunities for bacterial growth. However, the material is easily changed with static electricity which can draw dust when stored for long time. The temperature limitations are +100 °C and -40°C, but it must be realized that the material becomes brittle at low temperatures. Therefore it is not suitable as packing material for freezing.

Evaluation of physical and chemical properties of HD-polyethylene:

- Weight: The density of HD-polyethylene i8 low compared with many other materials.
- Corrosion: The material is resistant to rust and corrosion.
- Insulation effect: HD-polyethylene 15 a poor conductor of heat and provides a better insulation than metals.
- Workability- fabrication: HD-polyethylene is adapted to mass production methods and can be produced in large quantities at low manufacturing costs. On the other hand complexity of production methods reduces the opportunity to produce fish boxes without professional skills and equipment.

- Noise: The material makes some noise when it is handled but is less noisy than some materials such as metals.
- Toxity and resistance to chemicals: The material is not toxic and transfers no smell or taste to the contents. It is resistant to oil and grease and cleaning chemical.
- Surface: The surface is smooth and easy to clean.
- Colours: Products in a wide range of colours are available.
- Strength: HD-polyethylene has only medium strength and has disadvantages with respect to mechanical damage.
- Repairability: Broken boxes and containers are generally not repairable and must usually be replaced.

Aluminium Alloys

Fish boxes and containers made from aluminium alloys are expensive. They are manufactured in forms suitable both for stacking and nesting. and are well suited for freezing of fish.

Physical and chemical properties to be considered:

- Weight: Aluminium is known for its lightness compared with other metals (approx. one third of the weight of steel).
- Corrosion: Generally the corrosion resistance of aluminium is very good. Some foods react with aluminium. especially shellfish. (A protective lining/coating is recommended to prevent contact in such cases).
- Insulation effect: Aluminium has higher heat conductivity than other materials used for manufacturing fish boxes and containers. The insulation effect is therefore very poor. Faster freezing. however. is an advantage of aluminium.
- Workability-fabrication: Aluminium has good workability and is easily formed. Mass production is preferable.
- Strength: Aluminium boxes have a high mechanical strength.
- Toxity: Aluminium is non-toxic and is recommendable in direct contact with most foods but can reach with some shellfish.
- Surface: The surface is smooth. and is easily cleaned. although care must be taken with alkaline detergents and sanitizers.

- Repairability: Fish boxes and containers made from aluminium alloys are not easily mended. but in certain cases it is possible to repair them.
- Durability: Expected life is 8-10 years.
- Scrap value: When no longer fit for use. the aluminium fish boxes and containers have some scrap value.
- Noise: Fish boxes and containers made from aluminium are very noisy when handled empty.

Wood

Compared with modern materials as aluminium alloys and HD-polyethylene. wooden fish boxes and containers are rather cheap. -and are used mostly in less industrialized countries.

Wooden boxes can be used on vessels. in chill stores. in processing plants and in transport on trains and lorries. The capacity ranges from 20 to 150 kg.

Chemical and physical properties:

- Permeability: unsealed wood can absorb 15-20% water.
- Workability-fabrication: Fish boxes and containers made from wood can usually be produced near the plant in any quantity required.
- Repairability: Easy to repair.
- Toxity: Wood is a non-toxic material.
- Insulation effects: Wood gives a limited insulating effect which is lessened when wet.
- Surface: Wood has a surface that is difficult to clean thoroughly. Dealing with paint helps the cleaning. Unpainted wooden boxes are difficult to clean because of porous and rough surfaces.
- Durability: Wood has a short life-time (up to 1-2 years).
- Weight: The weight of the fish box will differ with the amount of water absorbed.

Disadvantages of Wooden Boxes

Difficult to clean. if handled roughly, they are easily damaged (mostly caused by inherent defects in the wood, poor design and poor nailing).

Expanded Polystyrene

Chemical and physical properties:

- Strength: The mechanical strength depends on the fusion process when made, but the material has less strength than other materials previously mentioned.
- Repairability: If the box is broken it cannot be repaired.
- Insulation effects: The material has superior insulation properties.
- Weight: The material is very light (down to 1/4 of wood).

Fibre Board

Boxes made of massive fibre board are primarily used as non returnable units. The fibre board is often coated with polyethylene inside and outside. These boxes are weak compared to wood, plastic and aluminium if subject to pressure.

List of properties to be considered:

- Strength: Boxes made from fibre board can sustain most types of handling except stacking without special arrangement.
- Insulation effects: The insulation properties of fibre board are comparable with plastic.
- Weight: The material is rather light (between wood and expanded polystyrene).

Table 3 Characteristics of various materials used in making fish boxes

	Stacking stability/strength	Relative density	Insulation	Durability	Repairability	Hygiene
Basket - Bamboo	Unstable/moderate	Light	Less good	Fair	Easy	Poor
Wooden box:						
-Unpainted	Strong/stable	Heavy	Good	Fair	Easy	Poor
-Painted	Strong/stable	Heavy	Good	Fair	Easy	Fair
HD - Plastic box	Strong/stable	Medium weight	Good	Good	(Imposs.) difficult	Good
Expanded polystyrene	Medium strong/stable	Extremely light	Extremely good	Bad	Impossible	Fair
Al. Alloys	Strong/stable	Medium weight	Bad	Good	Possible, but difficult	Good
Fibre board	Medium strong/stable	Light	Bad	Bad	Impossible	Good

Table 4 Physical properties of materials used in making fish boxes

Material properties	Strength (kg/mm²)		Heat conductivity (kcal/m h°C)	Density (approx. kg/dm³)
	Tensile strength	Bending strength		
I Box/container materials				
1 Aluminium alloy	20 - 30	30 - 40	160	2.7
2 GRP	20 - 50	30 -100	0.5	2.0
3 Polyethylene (HD)	5 - 10	13 - 15	0.35	0.94
4 Wood - Soft	5 - 8	8 - 12	0.15	0.5
5 Wood - Hard	8 - 14	10 - 15	0.20	0.7
II Insulation materials				
1 Cork			0.03	0.2
2 Polystyrene			0.027	0.35
3 Polyurethane			0.016 - 0.023	0.38
4 Mineral wood			0.03	0.2
5 Air			0.02	0.1

STANDARDS

The development of official standards is a key factor in creating the best possible adapted and cheapest product. Standardization of containers both nationally and internationally would be of great benefit to the fishing industry. This is difficult to achieve however, because the types of fishing and scale of operations differ from region to region, and to be effective containers must be suited to the fishery and must fit within the economic constraints of the industry. Despite these difficulties a great deal of mark has been done in many countries and several countries have developed national standards, often in cooperation with neighbouring countries. Work is also proceeding on the international scene and ISO (the International Standards Organization) has published over 4 000 recommendations and standards. The European-Standards Organization (CEN), with support from the EEC and EFTA countries and from Spain has published several standards.

THE INTENTION OF STANDARDIZATION

The advantages of standardization can shortly be summarized as follows:

1. production of packing materials can be rationalized by using the same equipment;
2. materials can be bought in large uniform quantities to a lower price;
3. the quality of raw materials and finished products can, to a greater extent, be controlled and improved;
4. it will be easier to cooperate on part deliveries from both raw material producers and production units;
5. several producers can specialize on partial production and/ or cooperation can take place as with regard to production of large quantities for the markets;
6. when different products reach the same processing operation, standardization will facilitate linking to the same processing and distribution lines;
7. reduce transport time and improve economy for freighted goods;
8. generally, standardization will result in economy for both producers and users.

Requirements for standardization of fish boxes have primarily come as a consequence of:

1. requirements by Health authorities for control of hygiene and quality of the raw material;
2. adaptation and standardization as a consequence of requirements by transport systems.

Standardization Systems

Scandinavia and parts of Europe, have approximately the same standards of fish production, a consequence of the trade in fish products. Within the different countries there is a great variation in sizes of boxes and containers. However, within Europe adaptation has taken place by using the measurements of the Euro pallet (80 x 120 cm). To keep the freight as low as possible the aim has been to stack boxes/containers optimally on pallets. Cold stores, lorries, railways and ships are often built according to the Euro pallet module. This will mean that area and volume of these forms of transport can be utilized to a maximum.

Marking and Safety of Goods

Marking of packing material must describe contents (fish species), weight, size of fish, sender and receiver. In international trade it is normally the receiving country that defines the standards of packing materials and the rules of marking. Safety of the goods is an important factor to consider when delivering to domestic as well as to export markets. Goods on pallets will sometimes slide off. This can be prevented by loading of uniform boxes in a stable regular pattern. The load can be further stabilized by the use of strapping or enclosing the whole load in shrink plastic film. Secure loading rules protect both the goods and the people handling them.

National Standards for Fish Boxes and Containers

If a standardization system has not been established it is important that each country try to adapt 4-5 different box sizes suitable for the following transport/handling and health requirements:

- should fit generally used transport pallets;
- suitable for transport by public transport systems such as ships, trains and lorries;
- they must be of a type which can be handled by jack trolleys and fork lifts;
- must be adapted to various loading and unloading equipment;
- should be made of hygienic and non-toxic material which is easy to clean.

We will now return boxes in general circulation in the fishing industry and used for transport of fish.

Exports

When fresh fish is exported from one country to another the exporter normally has to meet the requirements set by the importing country as to packing materials, marking, hygienic requirements, freshness of raw materials, etc. Success on export markets depends to a great degree on whether the exporter is able to adapt to standards set by the customer.

FISH BOXES - STANDARD REQUIREMENTS

Structural Requirements

Generally the requirements are the same for boxes used on board fishing vessels, in store, in production units and in transport on the road/railway. Boxes suitable for these uses are made of wood, aluminium or plastic. However, special requirements are set for boxes to be used in air freight transport, and it is especially important that such packing is waterproof.

The size of a box will be determined by its purpose, type of handling equipment available and the materials used on its construction. Boxes used inside a processing factory should be of a size suitable for one person to handle.

Boxes used in transport can be larger of loading equipment such as hoists and fans lifts are available at both ends of the trip. Standard sized boxes should be marked with a distinguishing number.

The Wooden Box

Dimensions and Accepted Deviation

The wooden box is generally made from soft-wood but hardwood can also be used although this tends to make it more heavy.

Table 5 Material thickness (pine and spruce softwood)

Box volume (in litres)		Material thickness (in mm)				
		Sideboard and Bottom board	End board	Stableboard and bottomboard	Corner block	Stiffener for topboard
From	To					
	25	10	15	10-13	15 x 38	10
26	45	13	15	10-13	15 x 38	10
46	80	13	15-20	10-13	20 x 50	10
81	150	15	20	10-13	20 x 50	10
151	300	18	25	10-13	31 x 62	10

Construction

Fish boxes must be made from dry materials which should have a plain and smooth surface to enable the best possible cleaning. Sides, bottom and top are assembled by nailing the boards together. In the bottom the boards are placed with an opening between each piece to enable water to drain from the box. The openings should not be bigger than 10 mm. Wooden boxes for reuse must have stiffening boards to obtain more strength. Boxes must be nailed with big-headed sharp edges nails of a length and a number that are dictated by material thickness.

Strength.

The boxes must stand normal daily loads. They shall without showing permanent deformation, stand an equally shared load corresponding to a 2.6 m high pile of boxes with 80% of volume filled with fish plus a load 3 times the weight of the pile.

The Plastic Box

Construction

The box must have a shape that utilizes the strength of the material and that gives stable stacking. The surface should be smooth, plain, without significant damage and have equally rounded corners. The box should be equipped with drain holes in the bottom or in the walls as near the bottom as possible.

Strength

The box should stand normal daily loads. It shall, without showing permanent deformation, stand a min. of 8 hours with a load corresponding to a 3 m high pile of boxes with 80% of the volume filled with fish and in addition a load 1.5 times the weight of the pile (short time test). The box should also stand long time loading test carried out under the same conditions as the short time test, but with a load 0.75 times the weight of the pile.

Durability against cracking

A plastic box should resist the following conditions: exposure to hot air at 100°C for seven days, submersion in a solution of 10% and synthetic detergents at 50°C.

Aluminium Alloy Boxes

Construction

The box should be made of AlMg2 aluminium alloy or of an aluminium material with at least the same properties and corrosion durability.

The box should be of a solid construction with the bottom, sides, and supporting flanges made from one piece of metal. The ends should be solidly jointed to the bottom and sides by smooth welding.

The box should be equipped with drain holes so arranged that the water and slime from a box is led to the outside of the boxes underneath it in a pile.

The surfaces of the box must be smooth, melds must be ground smooth and the box thoroughly washed out after manufacture.

Strength

The boxes should stand normal daily loads. They shall, without showing permanent deformation, stand an equally shared pressure corresponding to a 3 m high pile of boxes with 80% of the volume filled with fish and plus a load of twice the weight of the pile.

Fibre board boxes

The boxes are meant to be used as non-returnable packing material for:

- iced fresh fish;
- freezing of fresh fish;
- frozen fish

Construction

Fibre board material in boxes for iced or fresh fish should be smooth, free of cracks and dents, and humidity resistant on both sides. To obtain this the following treatment is normally used:

- polyethylene coating 10-20 g/m²
- high gloss waxing 10-25 g/m²
- wax coating 20-30 g/m²

If the boxes are glued, waterproof adhesive must be used.

In boxes used for freezing of fresh fish the inner side must be treated to a humidity resistance corresponding to a wax coating of 20 g/m^2.

The boxes material must not contain dangerous chemical or bacteriological substances which can be dissolved into the contents.

The boxes should be delivered flat from the factory and in forms that are easy to put together by nailing, glueing or using special closing mechanisms. The boxes can also be made of combinations of massive fibre board/wood or massive fibre board/ plastic.

Boxes for iced, fresh fish should be equipped with drainage holes in the bottom corners or in end walls as close to bottom as possible.

Strength

The boxes should stand normal daily loads.

Boxes for iced and fresh fish

The boxes shall without showing permanent deformation stand an equally shared load corresponding to a 1 m high pile of boxes with 80% of the volume filled with the goods in question, plus an additional load of 3 times the weight of the pile.

Special boxes in massive fibre board shall, without showing permanent deformation, stand an equally shared load corresponding to a 1.5 m high pile of boxes with 80% of the volume filled with the goods in question, plus an additional load of 3 times the weight of the pile. However, boxes with wood and plastic reinforcement when strapped should stand a 2.6 m high pile of boxes with 80% of the volume filled with the goods in question plus an additional load of 3 times the weight of the stack.

Boxes for freezing fresh fish

The boxes shall without showing permanent deformation, stand an equally shared load corresponding to a 1.5 m high pile of boxes with 80% of the volume filled with the goods in question, plus an additional load of 3 times the weight of the pile.

Expanded Polystyrene Boxes (isopor)

Construction

The material should be homogeneous without any dirty particles and discolouring.

The boxes should stand the use of common cleaning and sanitation treatments and washing with hot water. The box material must not contain dangerous chemical and bacteriological substances which can be dissolved into the contents.

Drain hales should be located along the sides and in the end walls as close as possible to the bottom. These hales should be of sufficient number and size (approx. 15 mm diameter) to allow complete removal of water and slime. The surface of the material should be plain and smooth with no jagged edges around the hales. Boxes meant for air freight transport should be watertight or equipped with water absorbent material.

Strength

The boxes should stand normal daily loads. They should, without showing permanent deformation, stand an equally shared load corresponding a 3 m high pile of boxes with 80% of the volume filled with fish, and in addition a load corresponding to the weight of the pile. They should also be able to withstand the impact of being drapped from a height of 3 m.

The load on the wall must not exceed 2 kg/cm^2.

Fish Boxes - Description and Area of Use

Returnable boxes

General

Within the world's fisheries there are great differences in vessel standards and sizes, fish species, methods of fish handling, etc. This means that a fish box will have to meet many different requirements so as to be suitable for use in different parts of the world. There are several types of boxes/containers that are used in the fishing industry, although only a few of these are designed and suitable for such use. Problems are encountered with the type of material, strength and construction/design. The design of returnable and non-returnable fish boxes is in most cases fairly similar. The difference is mostly found in type of material and method of construction (strength).

In the following some types of returnable boxes are described. The size and design characteristics apply mostly to boxes used in parts of the world with a relatively high technological level.

Wooden boxes

A selection of wooden boxes is shown in Table 6.

In developed countries labour costs for these types of wooden boxes are relatively high while the price will be quite favourable in most developing countries.

Box A is normally used for cod and fillets of haddock

Box B is used for small fish species like herring

Box C is used for fish of less than 5 kg like cod or haddock

Box D is the wooden box type that usually has been used to store whole fish on board fishing vessels and in production plants. The boxes have normally a solid construction (better than boxes A-C). The hygiene requirements imply that this type of box is usually painted

Box Eis used for bigger fish (20-30 kg), e. g., halibut and tuna

Aluminium Alloy Boxes

A selection of boxes is shown in Table 7.

These boxes are supplied in different shapes and sizes, and they are classified as returnable due to high price and durability.

Box A and B are suitable for internal use in processing plants when storing fish fillets and small fish

Box C is considered a suitable freezing mould and is normally used when freezing fish and fish products

Box D is suitable for use internally in factories to store fish and fish products. The facility nesting reduces storage space needed for empty boxes;

Box E is used for the similar purposes to box type D

Boxes of Polyethylene (plastic boxes)

A selection of boxes is shown in Table 8. The boxes can be ordered in several shapes and sizes and are made of HD-polyethylene (High Density polyethylene).

These boxes are classified as returnable due to high price and durability.

Box Ais used internally in plants to store fish fillets and small fish

Box Bhas similar uses to A and is sometimes used on fishing boats. The box is made of HD-polyethylene and can also be used for freezing fish and fish products. The design makes the boxes easy to nest when empty and stack when full;

Box Chas the same use as type D in aluminium. and it has also the same design. The box can be delivered with perforated bottom.

Box D and E have the same use as wooden boxes type D and aluminium alloy boxes type E

Non-returnable Fish Boxes

General

Non-returnable boxes are usually made from massive fibre board or styropor but may also be made from thin wood. Wooden boxes are not so widely used in industrialized countries due to the high cost level.

In developing countries wooden boxes will, in many cases, be most convenient to use and the cheapest alternative. The limitation will mainly be that wooden boxes are not waterproof and therefore unsuitable for air transport of fish without waterproof lining.

Boxes of fibre board and styropor are used to dispatch fish to wholesale and retail markets. These boxes are also the most commonly used when dispatching fish by air. The volume/weight designed to allow handling by one person, i.e., up to approx. 25 kg.

Some examples of practical uses of fibre board and styropor boxes for fish products follow.

Fibre board boxes

This box is delivered from the factory in 'knack down' form and erected prior to use by folding, stapling, stitching and glueing as required.

Fibre board boxes to be dispatched by air (Airenabox)

When air freighting fresh iced seafoods the main concerns are quality, safety and economy.

Two systems which satisfy the requirements of Scandinavian Airlines System (SAS) and other airlines will now be described.

The Airenabox is made of a special type solid fibre board with a very high water resistance because both surfaces are polyethylene lined.

The box has a telescopic design, and is water tight, even when turned upside down.

The instructions for assembly and use (stitching, taping and marking) must be followed closely as to get the box to function satisfactorily.

Table Dimensions of Airenabox fibre board boxes

Standard sizes and prices:

Internal dimensions:

Airenabox No	L x W x H (mm)	Volume (litres)
4415	385 x 385 x 150	22
6415	585 x 385 x 150	34
6422 a/	585 x 385 x 220	56
8323 b/	785 x 385 x 220	66
8323	isolated with foam lining	60
8425 c/	802 x 402 x 185	60

a/ No. 6422 is especially designed for lobster

b/ No. 8323 is especially designed for salmon

c/ No. 8425 is designed especially to fit into a standard expanded polystyrene box.

Table Prices of non-returnable boxes (approx.)

Type of box	Volume	US$
Fibre board box for fresh iced fish	38.5 litres	1.95
Fibre board box for salted fish/fillet	38.5 litres	1.70

Expanded polystyrene boxes (styrophor)

Boxes made of styrophor must be regarded as unsuitable for rough treatment and handling and are difficult to clean for re-use.

Because of good insulation qualities the box is suitable for transport of fresh iced fish as well as frozen products. When dispatched by car, ship or railway the box is used as it is, with the lid taped to the box. When dispatched by air sometimes the box is enclosed in a fibre board box. The standard box has dimensions L= 100, W= 400, H= 18 (mm) and will take 20 kg of fish and 10 kg of ice.

The above mentioned standard box costs with cover approximately US$ 1.85.

1. Flat packing material in store. Insignificant volume
2. The box (and lid) is stapled together, 2 staples in each corner as near the corners as possible.
3. Two side liner is placed in the box. The liner can be procured both with and without styrofoam. The styrofoam gives better insulation
4. An acceptable water absorbent pad is placed on the bottom plate of the box
5. A second bottom plate is placed on top of the absorbent pad. This to spread the pressure of the fish equally on the absorbent pad.
6. The fish is put in the box
7. Crushed ice is thrown directly on the fish. For extra insulation a top plate of styrofoam is used
8. A lid is finally put on and taped to the box, with approved water absorbent the box is acceptable for air transport without the tape. The box is finally sealed with tapes if requested.

1. Flat packing material in store. Insignificant volume.
2. Box (and lid) is stapled together, 2 staples in each corner as near the corners as possible.
3. An approved absorbent pad is placed on the bottom plate of the box.
4. A standard expanded polystyrene box is placed on top of the absorbent pad in the fibre board box.
5. The fish is put in the box.
6. Crushed ice is thrown directly on the fish.
7. Cover is placed on top of the polystyrene box.

- Alt. I A clear plastic or a massive fibre board cover is used.
- Alt. II A common polystyrene cover to suit the box is used.
- The reason why the polystyrene cover in Alt I is left out is that the box takes less space in height.

8. The lid is put on and taped into position.

This packing system is well suited for fresh as well as frozen products.

SPECIAL CONTAINERS

In the fishing industry there are a great number of containers in use designed for special purposes and uses.

They fall into two categories:

1. Insulated containers
2. Uninsulated containers

Generally their application is mainly within the following areas:

- storage (onboard vessels, in plant, with wholesaler, retailer)
- transport (sea, rail, road, air)

Choice of Containers

The main function of the container is to keep the fish in good condition while being stored or during transport.

Before deciding on type of container, special considerations should be given to the following points:

- main aim of using container;
- type of fish which will be held in the container;
- average storage time;
- type of storage (frozen, chilled, dry, etc.);
- means of handling and transport of the container;
- economic considerations.

The important qualities to consider when choosing a container are durability, suitability for the intended purpose and standardization for mechanized handling and transport. Long life and versatility will ensure maximum use of such equipment.

Insulated containers, by keeping out unwanted heat will preserve the quality of chilled or frozen fish. The use of appropriate ancillary equipment will facilitate handling, cleaning and maintenance.

The technology is available for manufacture of simple containers and also complex refrigerated units. Many types are in use around the world and a few of these will be described here. Those listed are practical, economical and cover most needs.

Standards for Simple Containers

The standards used are often a result of cooperation between the producer and consumer. The general requirements for chemical/ physical properties of the materials are that they should be non-toxic and easy to clean.

In Scandinavia and parts of Europe the base measurements used are often similar to the Euro-pall Modul, which is 800 x 1 200 mm^2. This size is easy to accommodate on transport equipment for the best utilization of area/volume.

Wooden Containers

Wood is not a very suitable material to use because it easily absorbs and retains water which makes it heavier and difficult to clean unless the surface is sealed. Wooden containers suffer damage relatively easily with rough handling and transport.

One advantage of wood is that it is available in most countries throughout the world. It is relatively cheap and can be worked with simple tools by local craftsmen.

Table Dimensions of polyethylene containers

Type	Volume approx. litres	Dimensions approx. mm									Goods thickness	Price US$	Weight approx. (kg)
		L	W	H	L1	L2	W1	W2	H1	H2			
A	675	1 520	1 060	670	1 460	1 350	980	900	520	120	7	250	55
B	750	1 200	1 000	1 050	1 140	990	940	790	850	120	7	465	75
C	1 000	1 200	800	1 630	1 140	990	740	590	1 450	120	7	500	90

Containers of different shape and volume can be made and repaired locally. Wood has a relatively good insulation effect, but is fairly dense.

Aluminium containers are solid and easy to clean but require specialized construction equipment and are more difficult to repair. Aluminium has poor insulation qualities. Containers made of seawater resistant aluminium will be quite expensive, but durable.

Steel containers are sometimes used for bulk fish transport. These are very durable if galvanized or painted with a rust proof paint. Sometimes steel containers have an inner GRP lining with the space between liner and steel filled with polyurethane foam.

Glass reinforced plastic (GRP) containers are simple to use and durable. GRP containers can be built to any shape and volume. The technology is much the same as used for GRP boat production.

HD-polyethylene is the material which is most commonly used for containers. This type can be double walled with air as insulation or have the hollow space filled with polyurethane foam. This "sandwich" construction gives good insulation and a stronger container. Insulation. Most insulated containers will have polyurethane or similar foam as insulation material. It has high insulation effect and is very durable.

Containers for Road Transport

The basis for chill/cold container road transport is the lorry chassis. These types of containers are insulated and built for fast fixing to/removing from the chassis.

Containers may be Built to Satisfy any Carrying Capacity

The load carrying capacity is of course not only dependent on the size of the container; due consideration must also be given to the capacity and stability of the vehicle, loaded and unloaded. It is also necessary that engine power and braking power are compatible with total weight of loaded vehicle and in accordance with international and/or domestic regulations.

The container is made of insulated laminates fixed to an aluminium profile bearing system for walls and roof. The bottom frame is made of steel profiles costomized to fit the lorry chassis in question.

Tested and approved steel fittings should be used for fixing the container's bottom frame to the chassis. It is the manufacturer's responsibility to fulfill this requirements.

Maximum permissible outside width is in many countries limited to 2 500 mm.

Table Standard measurements for lorry containers (mm)

Type	A	B	C	D	E	F	G
Inside width a/	2 300	2 360	2 430	2 430	2 430	2 360	2 220
Inside height b/	2 400	2 450	2 800	2 750	2 750	2 100	2 000
Lenght	adapted to the chassis						3 7004
							3005
							300

a/ Aluminium floor plate with raised edges reduces the inside width by 10 mm

b/ Has to do with max. standard measurements. Larger inside height can be produced on request.

Insulation

The majority of containers for modern road transport are made of pre-fabricated sandwich elements, usually hard polyurethane (PUR) foam laminated between sheets of aluminium or gel-coat finished GRP.

Doors are made of PUR insulated elements cast in one piece with cast-in reinforcement for fixing of hinges, locks. etc.

The floor is normally reinforced with an anti-slip aluminium sheets, bent up 200 mm along the sides and front wall. On top of a 15-21 mm plywood flooring.

Container Standards

On an international level there is a growing trend toward standardizing of structural details, surface finish, insulation and dimensions for food transport containers.

To optimize quality preserving conditions for perishable food-stuffs during transport, particularly in international commerce, discussions have been held at international levels. This has resulted in an agreement called "Agreement on international transport of easily perishable food-stuffs - and special equipment for such transport" (ATP).

The ATP agreement was developed in order to give transport container manufacturers a Code of Practice, or a norm for the design of their products.

Thus it would also be possible to classify the containers with respect to standards and suitability for various transport needs.

The transport container can be insulated only, with its own refrigeration equipment or insulated and heated. They are classified in different categories according to capacity, cooling system, K-value (measure for = insulation property), etc.

Testing of Containers

In order that new equipment may be approved for use, the ATP requires the equipment to be tested. Approval of a tested model or type is valid for three years, or for 100 container units.

Full tests for certificate of approval can be carried out in Germany at:

TECHNISCHER UBERWACHNUNGS VEREIN BAYERN EV. MUNIK (TÜV).

in France at:

CENTRE TECHNIQUE DU GENIE RURAL DES EAUX ET DES FROIDS, ANTONY.

Certificate

New containers that are already approved through tests on a "prototype" model, are awarded a performance certificate valid for 6 years. For later tests and tests on used containers, the validity of the performance certificate may be extended for 3 years at a time if the test is passed.

Marking of equipment Tested and Passed in Accordance with ATP Regulations Marking shall be in accordance with the regulations Dark blue Latin lettering on a white background shall be used. The minimum height of the lettering being 120 mm. The marking shall be easily visible, preferably fixed high up on the forward ends of both side walls.

Class	Equipment	Marking
	Normal insulation	IN
	Extra thick insulation	IR
Class A	Normal insulation cooled	RNA
Class A	Extra thick insulation. cooled	RRA
Class B	Extra thick insulation. cooled	RRB
Class C	Extra thick insulation. cooled	RRC
Class A	Normal insulation. mechanically cooled	FNA
Class A	Extra thick insulation. mechanically cooled	FRA
Class B	Normal insulation, mechanically cooled	FNB*
Class B	Extra thick insulation, mechanically cooled	FRB
Class C	Normal insulation, mechanically cooled	FNC*
Class C	Extra thick insulation, mechanically cooled	FRC
Class D	Normal insulation, mechanically cooled	FND
Class D	Extra thick insulation, mechanically cooled	FRD
Class E	Normal insulation, mechanically cooled	FNE
Class E	Extra thick insulation, mechanically cooled	FRE
Class F	Normal insulation, mechanically cooled	FNF*
Class F	Extra thi 'k insulation, mechanically cooled	FRF
Class A	Heated, with normal insulation	CNA
Class A	Heated, with extra thick insulation	CRA
Class B	Heated, with extra thick insulation	CRB

If the equipment is fitted with a removable temperature sensitive control unit, the asterisk "*" is added after the mark.

The certificate expiry month and year, is indicated immediately below the marking.

Example

FRC

05-1983

FRC denotes a container with extra thick insulation, mechanically cooled with temperature choice ranging between +12°C and -20°C. 05-1983 indicates that the equipment certificate expires May 1983.

The marking letter code is as follows:

The first letter indicates cooling (or heating) and type of cooling agent:

I - no cooling agent used

R - cooling agent is eutectic plates or liquid (Nitrogen)

F - mechanical cooling is used

C - heating can be (is) applied

The second letter indicates the highest permissible K-value:

N - normal insulation with highest permissible K-value = 0,7 W/m^2 °C

R - extra thick insulation with highest permissible K-value = 0,4 W/m^2 °C

The third letter indicates class and internal temperature range:

a. cooled equipment at external temperature +30°c

Class A +12 °C to 0 °C

Class B +12 °C to -10 °C

Class C +12 °C to -20 °C

Class D less or equal to +2°C

Class E less or equal to -10 °C

Class F less or equal to -20 °C

b. heated equipment capable of keeping internal temperature at least +12 °C when

Class A outside temperature is -10 °C

Class B outside temperature is -20 °C

Calculation of Required Capacity for Refrigeration Unit

The calculation formula used is

Q = A x t x k x 1.75 where

Q= required cooling capacity (W/h)

A=total area of roof, floor and walls/doors, measured at the middle of the insulated elements (m^2)

t= temperature difference between inside and outside of container (°C). (Inside temp. -20°C and outside temp. +30°C yields t= 50°C)

k= average k-value for the container as a whole (W/m^2 °C)

1.75= safety factor yielding an excess cooling capacity of 75%

once the required cooling capacity (Q) needed for the container in question has been computed, one must choose a refrigeration unit with cooling capacity equal to, or larger than this.

CALCULATION OF CAPACITIES AND CONSUMPTION OF ICE DURING HANDLING AND TRANSPORTATION

Ice is used in fisheries to chill the fish from surrounding temperature level down to 0°C and to keep it at this temperature. The weight of ice needed to chill 1 kg fish (0°C) can be calculated theoretically as shown below (in practice Some more ice will be needed):

Table Weight of ice needed to chill 1 kg fish from various ambient temperatures

Starting temperature of fish (°C)	Weight of ice needed (kg)
30°	0.34
25°	0.28
20°	0.23
15°	0.17
10°	0.12
5°	0.06

Ambient Temperature

The necessary quantity of ice required to maintain the fish chilled will depend upon the ambient temperature, the insulative properties of the container, the place of the individual box within the load and the length of the storage. In the following is given an example of ice requirements to chill and maintain in chill condition fish held in individual boxes and within a stack of boxes.

Table Ice requirements for chilling and storage of fish

	Melting of ice per box of 50 kg fish					
	1 box			35 boxes		
Surrounding temperature (°C)	+30	+20	+10	+30	+20	+10
Chilling fish (kg)	21	14	7	21	14	7
Keeping chilled (kg/h)	3	2	1	1	0.7	0.3

For practical purposes the following rules of thumb can be given to calculate ice requirements:

1. Fish boxes: Ice to fish ratio in tropics are 1 kg ice to 1 kg fish, and ice to fish ratio in temperate climate and in insulated van are 1 kg ice to 2 kg fish.
2. Insulated tanks: Water to ice to fish ratio in tropics are 1 kg water to 2 kg ice to 6 kg fish and in temperate climate 1 kg water to 1 kg ice to 4 kg fish.

Necessary volume of ice to chill the fish down to a temperature of 0°C is included in the above mentioned rules. If the fish is already chilled the volume of ice can be reduced accordingly.

Capacity of Boxes and Packing Density

The fish carrying capacities of various boxes and containers depend on the density of the mixture of ice and fish.

Table Density of different types of ice

Type of ice	Bulk weight kg/dm^3 = 1	Specific volumem^3/ton
Crushed block	0.690	1.45
Tube	0.565	1.80
Plate	0.570	1.75
Flake	0.445	2.25

Assumed that density of fish is 0.95 kg per dm³ (1 litre) the density of 1 dm³ (1 litre) of the mixtures of ice and fish in fish boxes, and water, ice and fish in tanks, areas follows:

Table Fish to ice ratio using different types of ice

Type of ice	Ice to fish ratio		Water to ice to fish ratio	
	1:1	1:2	1:2:6	1:1:4
Crushed block	0.82 kg	0.86 kg	0.90 kg	0.92 kg
Tube ice	0.75 kg	0.82 kg	0.87 kg	0.89 kg
Plate ice	0.76 kg	0.82 kg	0.87 kg	0.90 kg
Flake ice	0.70 kg	0.78 kg	0.84 kg	0.87 kg

From this table the capacity to carry ice and fish in different boxes and tanks can be calculated.

Table Icing using different types of boxes in temperate and tropical climates

Type of box	Volume (litres)	Tropics		Temperate climate	
		kg ice	kg fish	kg ice	kg fish
A	26	10.7	10.7	7.5	15.0
B	22 9	0 9	0 6	3 12	6
C	42	17.2	17.2	12.0	24.0
D	70 28	7 28	7 20	1 40	2
E	90	36.9	36.9	25.8	51.6

For an insulated tank with a volume of 1 000 litres, the similar calculation gives (crushed block ice).

Table Fish to ice ratio in an insulated tank in temperate and tropical climates

Tropics			Temperate climate		
water	ice	fish	water	ice	fish
100 kg	200 kg	600 kg	153.3 kg	153.3 kg	613.4 kg

HYGIENE REQUIREMENTS FOR DIFFERENT PACKING MATERIALS

Boxes, containers and other equipment which come in contact with fish during catch, transport and production, have to be cleaned as often as possible and at least in between each use.

Besides the natural micro flora found on fish and shellfish, external microorganisms from the surroundings are added. Bacteria and fungi come through the soil. the air and the water as well as from human contact with raw material.

Slime from fish provides a growth medium for most microorganisms. When the fish is dead, all defence mechanisms stop, and as a result the microorganisms living in the slime start to grow and attack the skin and the flesh. Offensive odours and flavours, discoloration and poor textures is the final result.

The rate of spoilage can be reduced by quick cooling and careful handling of the raw material and use of clean equipment which minimize bacterial contamination.

Water and Water Quality

Water is the cheapest and most convenient cleaning agent to use, but dried organic soil is difficult to remove by simply scrubbing with water, so detergents are added to dislodge, disperse and dissolve soil.

The quality of water used in the fishing industry is an important consideration because both chemical and bacterial contamination of seafoods can occur if polluted water is used. The level of dissolved minerals can also affect the solubility of cleaning and sanitizing agents in water.

Hard water contains elevated levels of calcium and magnesium ions. This water usually comes from limestone areas or from deep wells. The ions in hard water complex out detergents and so either a water softener must be added to remove the ions or excessive amounts of detergents will be needed for effective cleaning.

Rainwater and water from lakes and streams is usually softer, containing less mineral salts

Contamination of lakes and rivers from various pollution sources can render it unsatisfactory for use. In fishing, this water is used not only for cleaning but also for ice manufacture and direct human consumption, so its quality must be known.

Requirements to Water Quality

Bacteriological requirements

- Total acceptable plate count, max. 50 bacteria per ml;
- if there are between 2-23 coliform bacteria per 100 ml, the water and water source must be given a detailed examination before being used;
- water with more than 23 coliforms per 100 ml is unsuitable for use.

Physical requirements

- Colour = without colour (max. 20 mg Pt/l)
- Smell and taste = none, or very little
- appearance = clear and without sediments (max. 10 mg SiO_2/l)

Chemical requirements

- Permanganate-numbers (indicates content of organic substances) max. 20 mg kMnO4/l
- pH (acidity) 6-8.0
- conductance, max. 300 microsiemen
- sulphide, max. 0.1 mg/l
- Manganese, max. 0.05 mg/l
- copper. max. 0.05 mg/l
- hardness, preferably below 2° dH

These are standard values for a good water quality used for cleaning. Chlorination, filtration, and softening agents can improve water quality.

Detergents

General

A great number of detergents are found in the market. These substances are often complex and contain many different chemical components.

Most of the industrial detergents are corrosive and can, when used in a wrong way, cause damage to equipment as well as to persons.

The buyer of detergents has to make sure that the producer gives a thorough description of the chemical composition, areas of use and the correct dosing. Information about storing, safety requirements and marking as well as information on measures to be taken in case of accidents should be included.

Requirements to detergents

The cleaning solution must remove all dirt from the item being cleaned, without damaging it, even after it has been cleaned several times.

The detergent must be stable for storage, i.e., not change appearance form or quality when stored and be quickly and completely dissolvable.

Liquid detergents are easy to use, but expensive. Powder detergents are cheaper, but require more time for accurate dosing and dissolving.

The Cleaning Process

The cleaning process can be divided into two:

1. Washing
2. Desinfection

It is not possible to use these two processes in combination and get a satisfactory result. This is clearly illustrated in the following example, from tests in the food technology industry:

Table Effect of cleaning systems on bacteria

	Bacteria numbers/cm²
Before cleaning	2 900 000
After cleaning without detergent	40 000
After washing with detergent	5 500
After desinfection	10
Desinfection without cleaning	1 200 000

A well prepared washing programme has to include the following processes:

1. Mechanical removal of dirt (brush or other equipment)
2. Rinsing with tempered (or cold) water
3. Washing with hot water (40-70°C) with detergent
4. Rinsing with tempered/cold water
5. Desinfection of the clean surface leaning time for the sanitizer to work

Even if a cleaning programme can be set up in general terms, the variations will be many because of types of contamination to be removed, type of equipment to be cleaned and available cleaning equipment.

Cleaning methods

Because dried soil is very difficult to remove, hasing and mechanical scrubbling should be performed as soon as possible before the soil dries. This will reduce the effort needed later on.

There are many methods which can be used, and a satisfactory result can be obtained in several ways. According to the size of the equipment to be cleaned, the location and the frequency of cleaning needed, the following methods can be used:

1. Manual cleaning
2. High pressure washing
3. Foam wash

Manual Cleaning

Often, the working area is small or overloaded with other equipment so that manual cleaning is the only applicable method. Brushes of various types are simple to use and will give a satisfactory result in most cases. It is important that the equipment is kept clean and not used for other work.

This simple equipment is low priced and well suited for small operations or when the labour cost is low.

If dirt has been allowed to dry, the equipment to be cleaned should be put in water to soak for some time. Washing by hand will then be more simple and will in most cases give a good result.

Foam washing

This process is often used in the food industry. Special foaming detergents are sprayed onto the surface to be cleaned and left in contact with that surface to allow the detergent to dissolve the soil. This is particularly useful on processing equipment where the complexity of design makes manual cleaning impossible. The foam and soil is removed by hasing or manual cleaning. Unfortunately the "airy" consistency of the foam means that it soon assumes the temperature of its surroundings so the extra cleaning power of hot detergent is not possible using this method.

To make the foam, special equipment is needed. This is connected to a tank with special chemicals and the foam/water mixture is sprayed on the items to be cleaned.

High pressure washing

Equipment for this process is found in different shapes and sizes of both portable and stationary types. The small portable units are often adapted to small production units and are self contained.

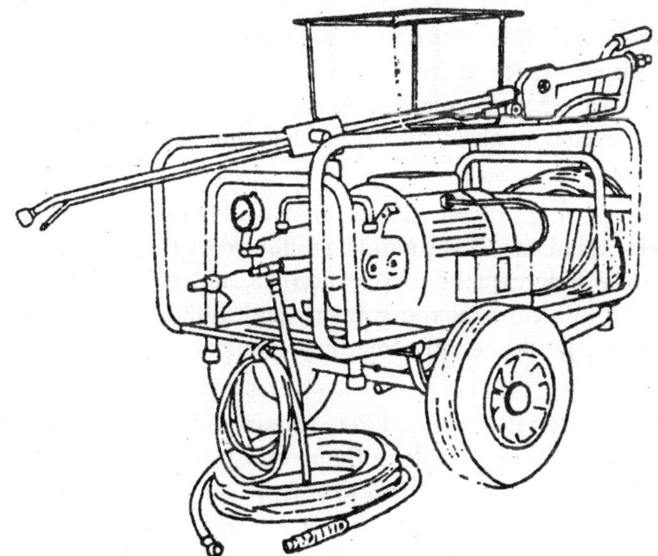

Figure High pressure washer

Total weight:	67 kg
Working pressure:	20.130 bar
Water capacity:	13 l/min
Water connection:	3/4 in hose
E1. Motor:	3 kW 220/380 V-3 phases
E1. cable:	8 m
H.P. hose:	8 m
Price:	approx. US$ 1 000

The stationary ones will be connected to pipe lines leading to strategic places. These units usually supply hot detergent water at pressure. The small portable units give up to 40-110 litre liquid per minute and work under a pressure of 40 kg/m². Some of these units have mechanical equipment for dosing and mixture of detergents and water. The large stationary pumps can give up to 300-litre liquid per minute under a pressure of approx. 55 kg/cm². When using such equipment the detergent can be dosed/mixed manually or automatically.

Desinfection

Washing and desinfection must be performed separately for maximum effect. There are two reasons for this. Detergents and sanitizers are often chemically incompatible and most sanitizers are complexed not only by microorganisms but also by any organic soil present.

The washing process should remove all the soil and some of the microorganisms. This should be followed by thorough rinsing to remove all detergent then application of a sanitizer or disinfectant which will kill most microorganisms and inhibit the growth of others. This should be left some time to allow it to work before rinsing off.

Desinfection is not only done to boxes and containers, but to all other equipment as well as floors and walls.

Below follows a rough list of existing types of desinfection products.

1. Chloride products
   a. hypochlorite (10-20 ppm/litre of water)
   b. chlorine amines
   c. clorisocyanur acids
2. Jodine products
3. Phenoles
4. Alcohols
5. Quaternary ammonium compounds

A variety of methods are available for application of sanitizers; these include manual spreading, spraying and fogging for enclosed tanks containers, etc..

Cleaned boxes made of modern materials like plastic, ought to have no more than 10.20 nos of bacteria per cm^2 on the surface when being examined.

DETERMINING QUANTITY REQUIREMENTS FOR FISH BOXES AND CONTAINERS

Quantity Requirements for Returnable Fish Boxes

The number of returnable fish boxes required in a box-pool depends on the marketing and distribution system. Many systems exist., from the most simple where fishermen retail their own fish to complex systems with many parties involved.

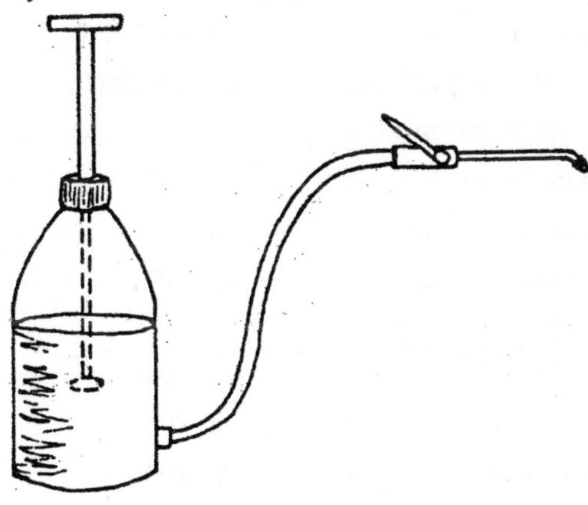

Figure Hand pump for desinfection

Two management systems are commonly used to distribute fish boxes to a fleet:

- the vessels draw from a special pool and operate with boxes provided from one place;
- the vessels deliver their catches to different receiving stations and exchange boxes when fish is unloaded. Fish boxes may be owned by the vessels in these cases.

A simple pool system is shown below. Fishing vessels or other fish transporting vessels land their catches to a processing plant or a market where fish is sold without bringing the boxes any further.

ECONOMICS

Production Cost

Costs of boxes and containers vary along with the level of cost in the various countries. If the costs are divided into material cost, production cost and overhead and profit, these will vary, but proportionally. The cost of materials and production machinery will be interrelated in the same way. Those proportional costs which are shown below (in percentages) represent an average from high cost countries (industrial countries). In Chapters 5 and 6 there are given some examples of retail prices.

Boxes: Proportional costs will be as follows:

a. HD-Polyethylene.

Cost of material	ca. 50%
Production cost	ca. 20-30%
Overhead and profit	ca. 20-30%

b. Aluminium alloy

Material cost	ca. 55%
Production cost	ca. 25%
Overhead and profit	ca. 20%

c. Wood

Material cost	ca. 70%
Production cost	ca. 20%
Overhead and profit	ca. 10%

The prices of wooden boxes will vary a great deal, depending on which country they are produced. However, it will be no problem calculating the prices if raw material and production costs are adjusted to domestic conditions.

d. Expanded polystyrene/solid fibre board.

The production process for this type of boxes is almost automatic. The costs are as follows:

Cost of raw material	ca. 55%
Production cost	ca. 10%
Overhead and profit	ca. 35%

Containers

a. HD-Polyethylene container for fish and ice.

Insulated with expanded polystyrene foam between inner and outer skin.

Production costs:

Cost of raw material	ca. 60%
Production cost	ca. 10-20%
Overhead and profit	ca. 20-30%

b. Glass reinforced plastic containers (GRP).

Insulated with expanded polystyrene foam between inner and outer skin.

Cost of raw material	ca. 40%
Production cost	ca. 40%
Overhead and profit	ca. 20%

c. Insulated containers for transport by trailer, i.e., 10, 20 and 40 feet.

The container is built with insulation of expanded polystyrene and covered by steel, aluminium or fibreglass.

The prices show small variations and can be estimated as follows:

Insulated for cooling	ca. US$ 275/m^3 1/
Insulated for freezing	ca. US$ 400/m^3

1/ Net volume of container

Production costs can be indicated as follows:

Cost of elements	ca. 45%
Production cost	ca. 45%
Overhead and profit	ca. 10%

Distribution Costs - Freight

Internationally there are few fixed freight rates for transportation of fish. Most often special freight agreements are negotiated in almost all types of transportation of fish and these depend on:

- Country and quantity per year
- Possibility of return cargo
- Price competition using same type of transport and using different types of transport
- Requirements for transport time

Lorry freight

The following price example shows estimated transport cost with lorry with a basis in price per 100 kg up to a volume of 3.5 m^3:

1 500 km	US$ 0.37/100 kg
2 500 km	US$ 0.50/100 kg
2 800 km	US$ 0.56/100 kg

The prices are for cold-stored and refrigerated container cargo, which are usually 20-50% higher than the rates of general cargo. For full containers the price will be lower. It is shown in the following example:

Full containers, i.e., 2 to 20 ft containers of 10 t with frozen or iced. fish can be transported a distance of:

- approx. 4 500 km for approx. US$ 5 500
- approx. 5 100 km for approx. US$ 6 000

Railroad freight

For railroad freight the prices are based on fully loaded containers with cold-stored or refrigerated fish.

Table 21 Cost structure in railroad freight

Distance (km)	Containers (US$)		
	10 ft	20 ft	40 ft
1 860			
2 800			
3 600	1 655		
2 065			
2 895	2 065		
2 480			
3 170	2 480		
3 310			
3 860			

As a basis of calculation for the cargo the following can be used:

10 ft - 5 000 kg; 20 ft - 10 000 kg; and 40 ft - 20 000 kg

Sea freight

The following table will give a basis for estimating sea freight: distance in nautical miles (1 842 m):

Table 22 Cost structure in sea freight

N. mile	120	300	500	1 000	1 400
kg	US$				
500	95	120	140	182	218
1 000	·193	235	291	361	417

In addition the following examples of freight rates:

40 ft cold-storage container equipped with refrig. unit and approx. 20 t of fish products.

Bangkok -USA East coast	US$ 5 800
Bangkok -USA West coast	US$ 4 800
Bangkok -Japan	US$ 3 000

Air freight

In the recent years it has become more and more usual to use air freight for transporting fresh fish and shellfish.

However, there are special requirements with regard to minimum weights if one wants to achieve the most favourable rates.

As an example freight calculation for fish/seafoods transported from West-Europe to USA are shown:

The minimum price per shipment is US$ 50

Special agreements are most often negotiated with regard to rate, especially for quantities above 100-500 kg. Containers are generally charged at a fixed price per container and weight. Price example from West Europe (US$).

FISH BOXES IN DEVELOPING COUNTRIES

In general in developing countries there is a confusing situation with a great variety of boxes and baskets being used in the fishing industry. This mixture of traditional local containers with modern plastic and aluminium boxes presents great difficulty for handling systems.

Traditional Boxes and Containers

Usually the availability of raw materials, quantities of fish landed and level of fishing operation are the factors which determine the types of containers which prevail in various regions/ countries. The following types of traditional containers are most common. but great variations occur:

- Leaf baskets (banana, palm tree, etc.)
- Fibre baskets (bamboo, ratten, etc.)
- Wooden boxes

Dimensions

These types of boxes/baskets come in different sizes varying from one area to another, but in general they will hold anything from around 1 kg of fish up to 60-70 kg. No standard measurements can be given.

Fig. Two types of traditional fish baskets used for fish (Africa and Asia)

Other Boxes/Containers

Boxes and containers of various types and materials are found in developing countries. The degree of use and standardization are decided by the importance of the fishing industry in the national economy and the direct contact with industrialized fishing nations.

In general containers used fall into the following groups:

- Boxes (plastic, aluminium, wooden, etc.)
- Baskets (plastic)
- Various (containers/tubs/buckets/trays of steel and wood and cut off oil drums. Different surface treatments are used e.g. galvanizing, paint and untreated),
- Insulated containers for special use (containers of various materials without/with insulation designed for use on vessels, processing units, trucks and fish markets/shops. Most of these are custom made, but standard types are available)

Dimensions

Sizes vary a great deal. but some plastic fish boxes follow the international standards (see Section 5.4), while in other cases any available box (bread, milk, etc.), will be used. Most uninsulated containers will take from 5 up to 70-80 kg, while the insulated ones range in size from a capacity of a few kg up to 20-30 t.

Suitability and Availability of Local Materials or Boxes and Baskets

Availability of local materials

The availability of raw materials for traditional baskets/boxes is in most cases sufficient as long as the demand does not increase dramatically. In some places the supply of the right type of wooden boards for boxes might cause problems, but this is more a planning and ordering problem than real shortage.

Suitability of local materials

Most local materials with the exception of wood can only be regarded as suitable for non-returnable or short lived returnable units.

The way these materials have been used and the traditional handling of boxes/baskets have created a reputation of the material not being hygienic. Also most traditional materials with the exception of wood are weaker and less durable than plastic and metals.

Since traditional materials most often are available within easy reach of box/container producers and the production quite often contributes significantly to the local industry, it is important that the use of traditional containers is not disregarded completely. Instead, with a more thorough approach to design and possible improvements, such units could be used with great advantages for the fishing industry. Improvements which could easily be achieved would be:

- More suitable shape and size
- Improvements on hygienic standards by applying sealants, lining and improved cleaning
- Improved production methods
- Increased strength

The only traditional material for which accurate information is available is wood.

In many places boxes and containers made from local materials are readily available in sufficient quantities and at reasonable prices, while modern fish boxes/baskets are costly and difficult to find. This means that if hygienic standards and other practical aspects could be improved on, the fishing industry would have less problems.

Spoilage Due to Lack of Adequate Packing and Handling Facilities

The rate of spoilage and quality deterioration is often quite high in most developing countries.

Spoilage losses range from 15% up to more than 30% and the two most significant contributing factors are poor handling and inadequate chilling.

Use of suitable boxes/baskets during the handling process and special containers for transport and storage will undoubtedly contribute to improved quality and reduced spoilage.

Improved fish handling methods should involve:

- Introduction of boxes/containers onboard vessels (if no other built in system exists), in the receiving system, and in the distribution chain
- Improved hygienic standards (proper cleaning and desinfection of boxes/containers)
- Use of ice or other cooling methods for fish stored in boxes/containers
- Improvements and standardization of boxes and containers in use

Introduction of Insulated Containers in Developing Countries

Insulated containers have a wide range of applications and come in various types from the more sophisticated types with refrigeration machinery to simple types used with ice.

The more advanced units are used in long distance transportation and international marketing. Generally it will be the transport distance and the capacity of the marketing industry which will decide the usefulness of such containers.

For the less advanced fishing industry, simple custom made insulated containers for iced fish will offer an inexpensive solution if one wants to introduce improved fish handling at critical stages. The following areas will achieve the greatest advantages from introduction of adapted insulated containers:

Small vessels: It is possible to arrange built-in or removable insulated containers in most small vessels. Care should be taken not to upset stability with such installations.

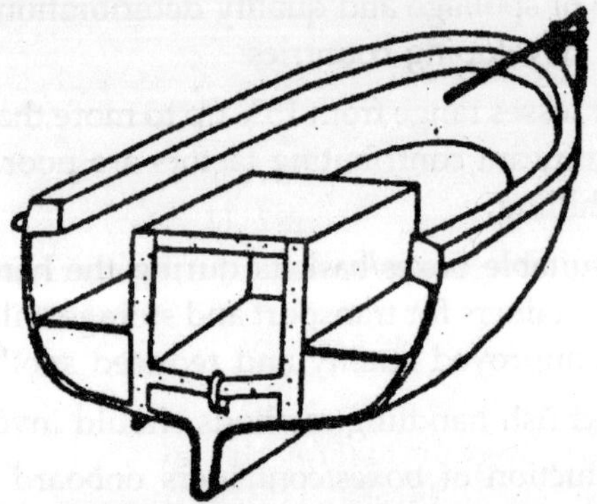

Figure Insulated hold in small vessel

Section of small fishing boat with insulated container for catch

Road transport

Insulated containers are widely used in conjunction with trucks, trains and ships, but in many countries a significant amount of fish is transported over short distances on bicycles and handcarts and often in uninsulated boxes/baskets. Light weight insulated containers can easily be made to suit such vehicles and would introduce a possibility for improvement in fish quality.

Retail market

Retail sale of fresh fish requires space for display and storage. Storage can be arranged by use of insulated containers with ice. This will offer an inexpensive alternative which can keep fish in excellent condition for more than a week.

Such equipment will be within financial reach of small-scale retail businessmen who cannot afford any other refrigerated equipment.

Simple containers used for iced storage must have a smooth accessible interior that can be easily cleaned.

The lid should also be insulated, well sealed and positioned in the top. It is important that the container has a drainage opening at the lowest point of bottom. 5-10 cm of styrofoam insulation material will give sufficient protection and GRP (Glass Reinforced Plastic), various types of plastic sheeting, aluminium, galvanized steel and wood (marine plywood and timber) can be used as construction material.

It is important that insulation material is distributed evenly through all parts of the box and that the inside which is going to be in direct contact with the fish is made, from hygienic and non-toxic material.

UNITS

The units used throughout the text are conventional metric units with a few additional. A British units.

For the convenience of the reader a table listing a few conversion factors for metric and British units used in the text are presented in the following table.

Selected International System Units

Power:	1 kilowatt (kW) = 1.341 hp
Specific energy:	1 kilojoule per kilogramme (kJ/kg) = 0.239 kcal/kg
Heat flow rate:	1 watt (W) = 0.86 kcal/h 1 watt (W) = 3.412 Btu/h
Specific heat capacity, mass basis:	1 joule per gramme degree Celsius (J/g deg C) = 0.239 kcal/kg deg C
Thermal conductivity:	1 watt metre per square metre degree Celsius (W m /m^2 deg C) = 0.86 kcal m/ m^2 h deg C 1 watt per metre degree Celsius (W/m deg C) = 0.86 kcal/m h deg C
Thermal conductance:	1 watt per square metre degree Celsius (W/m^2 deg C) = 0.86 kcal/m^2 h deg C

Table Conversion factors for British/SI units

Quantity	British units	SI units	Conversion factors	
Length	foot and inch	meter (m)	1 foot = 0.3048 m.	1m = 3.28 foot
		decimeter (dm)	1 foot = 3.048 dm.	1 dm = 0.33 foot
		centimeter (cm)	1 inch = 2.54 cm	1 cm = 0.39 inch
		millimeter (mm)	1 inch = 25.4 mm	1 mm = 0.039 inch
Area	square, yard, foot, inch	square meter (m^2)	1 yd ^{2}= 0.836 m^2	1 m^2 = 1.196 yd^2
		square decimeter (dm^2)	1 ft^2 = 0.093 m^2	1 m^2 = 10.76 ft^2
		square centimeter (cm^2)	1 $inch^2$ = 645.16 mm^2	1 mm^2 = 0.0016 $inch^2$
Volume	Cubic foot, inch	cubic meter (m^3)	1 ft^3 = 0.028 m^3	1 m^3 = 35.31 ft^3
		cubic decimeter (dm^3)	1 m^3 = 645.16 mm^3	1 mm^3 = 0.0016 m^3
Mass	ton, pound	metric ton (t)	1 ton = 1.016 t	1 t = 0.98 ton
		kilogram (kg)	1 lb = 0.454 kg	1 kg = 2.20 lb
		gram (g)	1 lb = 454 g	1 g = 0.002 lb
Volume, liquid	gallon	liter (l)	1 gal = 4.546 l	1 l = 0.22 gal

3

Fresh Fish Handling

INTRODUCTION

The Fishery Industries Division of the Food and Agriculture Organization of the United Nations is receiving numerous inquiries on specific data concerning planning, design and operation of fresh fish handling facilities from governments and Industries mainly of countries developing their fishery industries. In addition, requests for direct assistance in carrying out prefeasibility studies, occasionally of relatively simple nature, in the field of fresh fish handling are received, and in order to provide the assistance often highly specialized engineers have to be engaged since only they have access to some of the required basic engineering data. Also, for other feasibility studies and banking projects in this subject matter field the required engineering data are not always readily at hand.

As a matter of fact, the engineering data concerning fresh fish handling are either scattered in different publications or exist only in the files of individual engineers as they were never published. Considerable effort is in practice wasted by individual engineers to construct the required data for their own use, which results in both heavy duplication of work and often inaccurate data.

This publication is intended to assist field staff in the selection or specification, costing (both capital and running) and planning of facilities or equipment concerned with fresh fish handling and storage. It is not intended as a code of practice. It is hoped that it will help to satisfy requests for information more speedily, to contribute to improved accuracy and to save on efforts on parallel construction of data. This document does not pretend to include all the relevant data at this stage. It is recognized that a complete presentation would require time and resources which could not be justified for an initial effort, and that some data, such as costs are subject to fast changes. However, the intention of the document is to serve as a core guide to which additional data will be added by the users in the future. The users are, therefore, cordially requested to communicate their comments and all the additional data they find useful to the Fishery Industries Division. FAO, 00100 Rome, Italy, for later incorporation in an improved, more permanent edition of the document.

The publication is not the result of original research by the author but Is based on information available from manufacturers of equipment, from commercial plant operators and from previously published material by numerous organizations, particularly the Food and Agriculture Organization of the United Nations, Torry Research Station, Aberdeen, UK, and The White Fish Authority, Industrial Development Unit, Hull, UK.

It must be stressed most strongly that the costings given in the report are based on UK costs for 1979. Local costs of labour, land, services and costs of delivery and erection can significantly affect these costs. Given an appreciation of local costs relative to the UK, however, these figures can serve as a basis for budget costing and planning failing any better sources of information.

PHYSICAL DATA

Fish are commonly classified by physical characteristics or patterns of behaviour and are usually referred to as either pelagic or demersal. The pelagic fish such as mackerel and herring are feeders of microscopic plankton found in the surface layers of the sea. Demersal fish such as grouper, cod and flatfish, live on or near the sea bed and these patterns of behaviour give rise to different fishing techniques. Generally the pelagic species of fish contain higher levels of fat content than the demersal species.

Fatty Fish

Fatty fish contain most of the fat content in the body tissues and use it as a source of energy. The fat is concentrated in the flesh about the lateral line running down each side of the body and will vary considerably depending upon the time of the year. The herring for example (*Clupea harengus*), may have a fat content of only 1 percent immediately after spawning and more than 20 percent at other times when feeding well. The percentage increase in fat content is at the expense of the water content with the protein level remaining fairly constant. Figure 1 shows the seasonal variation of fat and water for the mackerel (*Scomber scombrus*) in the UK.

Lean Fish

Lean fish are typically bottom-living fish with low fat content. Usually less than 5 percent. The bulk of the fat is contained in the liver with the muscle having a low fat content.

Chemical Composition

The chemical composition of fish depends on the fish species, the season, the condition and the feed of the fish. Main components are water, fat and protein but the low concentrations of carbohydrates, minerals, vitamins, sugars and free amino acids also present are important to flavour, odour and nutritional value. Table 1 gives the composition of the raw flesh of a number of selected species (main components only).

Table 1 Chemical composition

Species	Scientific Name	Water (%)	Fat (%)	Protein (%)
Cod	*Gadus morhua*	78-83	0.1-0.9	15-19
Redfish	*Sebastes sp.*	73-79	3.2-8.1	16.8-19.7
Herring	*Clupea harengus*	60-80	0.4-22.0	16-19
Mackerel	*Scomber scombrus*	56-74	1.0-23.5	16-20
Hake	*Merluccius merluccius*	80	0.4-1.0	17.8-18.6
Sole	*Solea solea*	78	1.8	18.8
Albacore	*Thunnus alalunga*	59-72	4.3-16.1	21-27

Specific and Latent Heat

Specific and latent heat values of fish are dependent upon the chemical composition of the fish but for lean fish with a water content of about 80 percent, the specific heat is 0.9 kcal/kg°C for temperatures about -1°C, the temperature at which lean fish starts to freeze. As freezing does not take place at a particular temperature latent heat and specific heats are removed simultaneously.

The heat content of fatty fish, which has a lower water content than lean fish, will be reduced. A fish with a fat content of 20 percent may start to freeze at -2ºC and have a heat content lower by 22 percent. Since the oil content varies seasonally the figures for lean fish may be taken as representing the maximum cooling requirement.

Perishability

When a fish dies spoilage begins due to bacterial, enzymatic and chemical action. On death the bacteria present in the surface slime, gills and intestines, which do no harm to the living fish due to its natural resistance to them start to multiply and penetrate the tissues, the enzymes in the stomach and intestines start to penetrate the belly lining and oxidation of the fat in the flesh occurs.

The rate of spoilage is dependent upon the holding temperature and is greatly accelerated at higher temperatures, due to increased bacterial action at the higher temperatures. Freshness is to a certain degree subjective but it can be "measured" against an agreed scale by assessment of appearance, odour and-taste. Fatty fish will deteriorate far quicker than lean fish due to the higher rate of oxidation of the fat. Herring for example will be unacceptable after five or six days when held at 0ºC and after only 30 h when held at 15ºC. Spoilage can be reduced by washing, gutting, gilling and above all by keeping the fish cool by the use of ice.

Table 2 lists the shelf lives of selected species.

Table 2 Shelf life

Temperate Water		Tropical Water	
Fish	Shelf-life (days)	Fish	Shelf-life (days)
Cod	12-15	Snaper (Brazil)	11-16
Haddock	12-15	Tuna (USA)	29
Whiting	9-12	*Synagric japonicus* (India)	27
Hake	8-10	Bonga (West Africa)	20
Redfish	13-15	Sea bream (West Africa)	26
Herring	5-6	Burrito (West Africa)	22
Mackerel	7-9	*Tilapia* (West Africa)	28

Weight-length Relationship

The relationship between weight and length of fish may be of interest in the design of fish processing machinery and fish handling equipment.

Age-weight Relationship

The age-weight relationship of a fish is of particular interest in the husbandry and harvesting of farmed fish and for any given species will depend on the type of feed, the water temperature and quality, the holding method and sex of the fish.

ICE

Ice has long been used to cool, and thereby preserve fish both at sea and on-shore. It has a large cooling capacity for a given weight, is relatively cheap, prevents drying by keeping fish moist, is easily portable and maintains a temperature slightly above the freezing point of fish without the need for sophisticated temperature control. It can be made from fresh water or sea water which should in both cases meet micro-biological standards of potable water and be free of objectionable substances. Ice made from polluted water can contaminate fish and reduce its keeping time.

Physical Properties - Fresh Water Ice

The physical properties of fresh water ice are given in Table 3.

Table 3 Physical properties of fresh water ice

Property	Metric units
Melting point	0°C
Density at 0°C a/	0.92 t/m°
Specific heat at 0°C	0.49 Kcal/Kg °C
Specific heat at -20°C	0.46 Kcal/Kg °C
Latent heat	80 Kcal/Kg
Thermal conductivity at 0°	1.91 Kcal/mh °C
Thermal conductivity at -10°C	1.99 Kcal/mh °C
Thermal conductivity at -20°C	2.08 Kcal/mh °C

a/ The figure quoted for the density of ice should not be confused with the stowage density of different forms of ice which are given in Table 4

Table 4 Stowage densities of ice types

Type	Stowage (mýÿ/t) a/
Crushed block	1.4-1.5
Tube	1.6-2.0
Plate	1.7-1.8
Flake	2.3-2.3

a/ Note that these stowage densities In Table 4 do not necessarily relate to the size of store required for a given weight of ice without allowance being made for handling.

The cooling capacity of the different types of ice will be practically the same for equal weights but not for equal volumes due to the difference in stowage density. Subcooled ice will have marginally more cooling capacity due to being sub-cooled and dry. Wet ice caused by hot gas or water defrost systems will have slightly reduced cooling capacity due to the free water contained in the ice, but for practical purposes the cooling capacity of equal weights of ice types can be considered the same.

The cooling rate or rate of meltage of ice will not be the same for different types of ice and will depend on the surface area of the ice. For example, flake ice has a surface area of over four times that of tube, weight for weight which can be an advantage or a disadvantage depending on its intended application. A poor stowage rate is a disadvantage for applications where space is limited and will give a slightly reduced capacity of fish in a box. Flake ice also has a tendency to bridge over the fish leaving an air space between the fish and the ice, thereby reducing effective cooling but is easier to handle as it is dry and produces the least damage to fish of all the forms of ice.

Seawater Ice

Seawater is suitable for ice making provided it is not contaminated and produces an ice soft and wet in appearance. It is not stable and during storage the brine tends to leak out leaving fresh water ice. The freezing point is dependent on the salt content as shown in Figure 8. When used to cool fish partial freezing of the fish can occur which is not desirable and salt uptake in the fish flesh is noticeable with prolonged storage. Depending on salinity and the amount of entrapped air it has a density in the range of 0.86 to 0.92 t/m^3. It has application where fresh water is expensive or in short supply and can be used for production at sea.

ICE PLANTS

Types

There are many methods of producing ice and the selection of the most suitable plant for a prospective buyer is not simple. The manufacturing process of one may involve efficiencies or advantages over another or the ice produced by one process may be better suited to the intended application than another. The plant types are usually referred to by the type of ice they produce; common examples being block, flake, plate and tube.

Block Ice

In the manufacture of block ice, tapered rectangular cans filled with water are immersed in a tank of refrigerated brine as shown in Figure below.

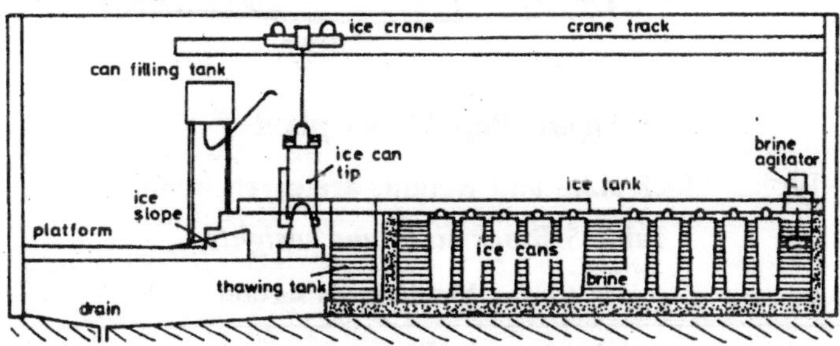

Figure. A brine type block ice plant

The brine which is cooled to about -5°C by a refrigeration process extracts the heat from the water and produces block ice within the can. The cans are then removed from the tank and thawed for a short time in a tank of water to release the block from the can. The blocks are then stored in a cold room and can be crushed on demand. The freezing period is typically between 16 and 24 h although plants known as rapid-block are available that have freezing periods of only a few hours. The quick freeze is achieved by the direct evaporation of the refrigerant in a jacketed mould fitted with finger evaporators. The blocks are released by a hot gas defrost and are subcooled to a temperature of -8°C. Rapid block plants require far less floor space than brine tank systems. Figure below shows a rapid block plant.

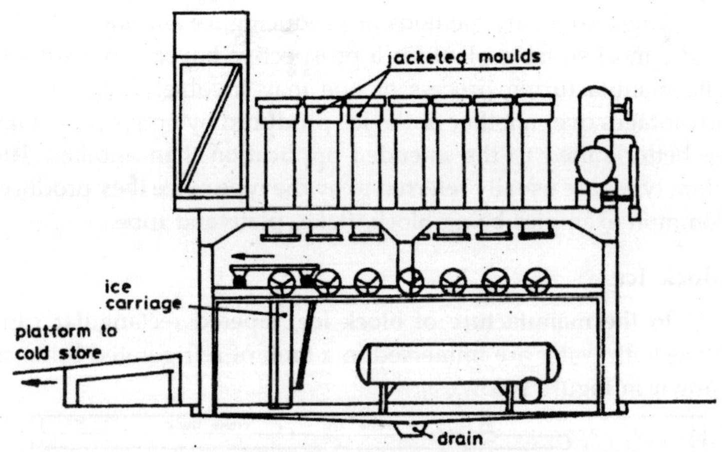

Figure. Rapid block plant

Typical block sizes and weights are given in Table 5.

Table 5 Block sizes and weights

Weight of block (kg)	Mould cross section Top (mm)	Bottom (mm)	Length (mm)
25	270 x 140	240 x 114	1 025
25	190 x 190	160 x 160	1 105
50	408 x 184	378 x 154	1 000
136	560 x 280	535 x 254	1 220

Flake Ice

Flake ice is formed by spraying water over the surface of a refrigerated drum to freeze it and then mechanically removing it with a blade as shown in Figure below overleaf. In some models the drum rotates against a stationary scraper on its outer surface; in others the scraper rotates and removes ice from the inner double-walled stationary drum (as shown). No water is sprayed on the drum immediately in front of the scraper, so that the water is completely frozen and dry on removal. Ice produced by this method is commonly 2-3 mm thick.

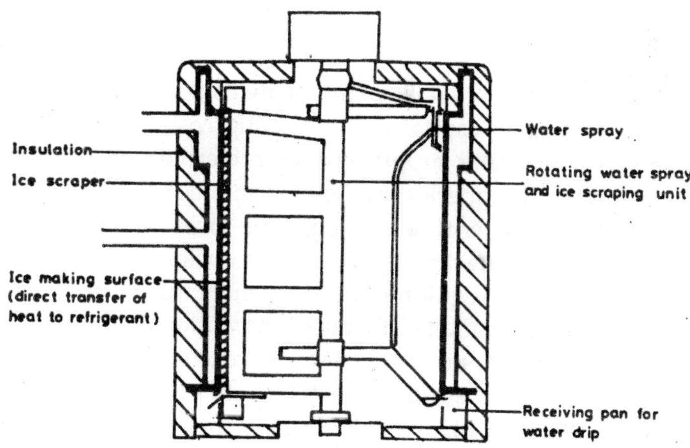

Figure. Flake ice plant

Plate Ice

In the manufacture of plate ice, water is sprayed and frozen onto the outer surface or surfaces of a refrigerated plate to form a sheet of ice which is usually released by an internal hot gas defrost.

Thickness of the ice can be varied, within limits, by alteration of the defrost cycle, but is commonly set between 8 and 15 mm. To produce ice thinner than this would be inefficient due to the refrigeration energy required to repeatedly account for the heat of defrost. As the ice falls it is chopped by a rotating cutter. The process is automatic.

Tube Ice

Tube ice is formed in a vertical shell and tube vessel by passing water down the Inside of tubes which are cooled by the circulation of refrigerant on the outer surfaces.

When the ice reaches the desired thickness the water flow is stopped automatically, the refrigerant is removed to a surge drum and hot gas is circulated around the tubes causing the ice tubes to melt and slide down. As the ice falls from the tubes it is broken by a rotating knife blade to a desired size. The ice tubes are about 50 mm in diameter with a wall.thickness of 10-12 mm.

Selection

In the selection of an ice plant, many considerations need to be made relating to capital costs, running costs, availability of skilled servicing, water supply and the preferred type of ice. No plant is universally superior to all others and the selection of plant needs to be made with knowledge of local conditions and requirements. Manufacturers will meet a customer requirement from a given range of equipment and can manipulate specification to suit a particular market or to meet competition. In the selection of equipment or comparison of competitive tenders, it is essential to consider the total costs of production; both running and capital. It is possible that a plant of low capital cost could have high running costs resulting in a higher cost of ice than a plant of higher initial cost but lower running costs. It is essential to fully understand and appreciate a manufacturers specification and exclusions.'

It is usual at the appraisal stage of an investment to prepare a technical/economic analysis based on estimated capital and running costs. The analysis should help not only in financial planning and pricing but in selection of plant as the relative local costs of electricity, water, finance and labour will influence the design specification.

As an example, the following analysis provides a costs estimate of ice production. The analysis is based on a flake ice factory of 150 t/24 h. The capital and running costs used are 1979 costs in the UK The analysis assumes a location on the dockside and includes no allowance for distribution.

A Cost Estimate of Ice Production

Flake ice plant - 150 t/24 h

Total years production = 120 t/24 h x 5 d x 52 weeks = 31 200 t (at 91% capacity utilization)

Costs/year	US$	% of annual cost
1. Labour (inc, of Soc.Sec. payments, etc.) 2. Electricity 3. Water 4. Office supplies 5. Depreciation (plant at 10 years, buildings at 15 years 6. Capital interest at 10% 7 Maintenance 8. Land, rent, taxes and insurance 9. Audit, legal and bank charges 10. Txes on profit 11. Dividend to shareholders 12. Miscellaneous (pension fund, bad debts, etc.) Total annual costs	50 000 68 000 8 000 6 000 83 000 100 000 20 000 10 000 3 000 20 000 24 000 3 000 395 000	12.7 17.0 2.0 1.5 21.0 25.3 5.1 2.5 0.8 5.1 6.1 0.8

	US$
Cost of ice/t	12.7/t
Assume sale price	17.0/t
Revenue from sales	530 400
Profit	135 400
Return of investment	13.54%

It is clear from the analysis that the correct selection of capacity of plant and store to meet the size and pattern of demand is of importance. If the plant is overspecified and underutilized the fixed overheads will be carried by a disproportionate volume of sales and the unit costs will be high.

The ice store should be sized according to both the volume and pattern of demands. Ice production and demand are seldom in phase and the store acts as a buffer helping to overcome daily fluctuations between production and demand. It also acts as a buffer in the event of a temporary breakdown of plant or interruption of services. Stores are commonly 2-4 times the daily production capacity. To assist the supplier or refrigeration contractor in selection and design, the buyer should make available as much information as possible regarding local costs and site conditions, etc. The following lists the more important considerations:

- Production capacity desired under local ambient conditions
- Ice storage capacity desired
- Purpose for which ice is to be used
- Preferred type of ice, and in the case of block, the block size
- Maximum ambient temperature and humidity and annual fluctuations
- Cost and availability of local labour including skilled
- Temperature, pressure and purity of ice make-up water
- Cost of make-up water
- Temperature, pressure and type of condenser cooling water
- Cost of cooling water
- Information on electricity supply: voltage, cycles, phase, maximum installed power, maximum starting current allowable. Cost of electricity/unit and details of any reduced rate for off-peak use.
- Detail of any physical or planning restrictions of the intended site
- Detail of site soil tests
- Particulars of grant aid, or soft loans (if any) toward capital or running costs of plant and conditions attached to aid

Cost of Capital (interest)

From the information provided selection of plant can be made that will satisfy any site restrictions and produce ice at a minimum cost.

Capital Costs

It is often necessary for purposes of financial planning that budget costs are known before a detailed analysis is available. To assist in financial and technical planning the following tabulated data is an attempt to provide an order of costs and physical data for different types of ice plant inclusive of store. The figures shown in Table 6 are based on information provided by plant operators in the UK and assume UK cost levels (1979) and ambient conditions. It must be stressed most strongly that these figures cannot be used universally without considerations to local site conditions and costs, but they do provide a datum from which rough budget costings may be estimated given a general appreciation of local costs compared with those of the UK.

Table 6 Capital costs (US$)

Type	**t/24 h**						
	2	**5**	**10**	**20**	**50**	**100**	**200**
Block (brine)			500 000	600 000	800 000	1 300 000	1 900 000
Tube	20 000	40 000	180 000	280 000	460 000	740 000	1 300 000
Flake	15 000	30 000	160 000	270 000	420 000	720 000	1 280 000
Plate			140 000	240 000	400 000	700 000	1 240 000

Automatic operation and conveying is assumed on production of 10 t/24 h and above for tube, flake and plate. Manual operation and discharge is assumed for 2 and 5 t/24 h. Three days storage capacity is included for tube, flake and plate and two days for block. The figures represent a turnkey cost and are inclusive of plant, erection, delivery, soil investigations, commissioning, freight and insurance, etc., but exclusive of ground cost. The costs reflect a site that requires no exceptional civil works in piling or clearing with services near at hand.

As a rule of thumb the plant items (ice generators, refrigeration machinery, motors, conveying equipment, weighing equipment, etc.) for tube, flake and plate ice, plants over 10 t/24 h, amount to approximately 50 percent of the total cost.

Area Requirements

The required overall ground area for an ice plant will largely depend on the type of: ice-maker, condenser and store and the configuration of these units. It is usual in the case of plate and flake ice production that the ice-maker be located over the store enabling gravity feed of the ice to the store which results in a low ground area requirement overall. This is not feasible for the block type of plant because of the heavy loads imposed on the structure and unusual in the case of a tube plant because it would result in an excessively high building. Table 7 shows the area requirements of the ice-maker and refrigeration plant, not including storage, office accommodation, electrical sub-station or condensers.

Table 7 Area requirements for ice-maker and refrigeration plant (m)

t	2	5	10	20	50	100	200
Block	-	-	100	200	450	800	1 500
Tube	-	-	30	35	60	92	160
Flake	7 a/	10 a/	20 a/	25	50	80	140
Plate	7 a/	8.5 a/	9 a/	12 a/	50	80	140

a/ Packaged units with allowance for access

Direct Running Costs

The direct running costs of labour, electricity, water and maintenance will largely depend on the type and size of plant, location and the degree of automation. As mentioned before, the design of plant should reflect the local capital and running costs such that the cost of ice per ton is minimized.

In some instances there will be a trade-off of component costs. For example, the selection and design of the condenser requires a knowledge of local costs of electricity, water and labour (ignoring space as a cost). A shell and tube condenser using sea water which is dumped would have low water and electric costs in comparison to other condensers but higher labour costs due to the cleaning required. An air-cooled condenser would have no associated water cost but higher electric cost (particularly if it was silenced). An evaporative condenser or shell and tube with cooling tower, using fresh water would compromise the costs of electricity and water.

Power

In addition to the type of ice plant and its associated refrigeration machinery, the power consumption will depend on local ambient conditions and feed water temperatures. Table 8 shows the approximate electricity consumption per ton of ice produced for the icemaker and refrigeration plant for temperate and tropical areas. The figures do not include requirements for handling, crushing or storage.

Table 8 Approximate electricity consumption for ice production (kWh/t)

Type of ice	Climatic Zone	
	Temperate	Tropical
Flake	50-60	70-85
Plate	45-55	60-75
Tube	45-55	60-75
Block	40-50	55-70

The flake ice plant has a slightly higher overall requirement due to the low evaporation temperature used to produce subcooled ice even though it has no defrost load requirement. Large ice plants are often more efficient than smaller ones and are sometimes specified with precoolers particularly in tropical areas, which sometimes enables the use of a smaller icemaker. The precooler is specially designed to lower the temperature of the feed water to the ice-maker and does it more efficiently than the ice-maker. Automatic plants (flake, tube and plate) are better suited to take advantage of any tariff reductions for off-peak use of electricity. Where penalties are incurred for high peak demands such that occur on start-up, the plant can be supplied with a current-limiting device that will limit the power demanded.

Labour

Labour costs will depend on local rates of pay, social security payments, the degree of automation and the type and size of plant. Generally the automatic plants (flake, tube, plate) require far less labour than block plants, particularly for the larger capacities. The cycle and harvest of automatic plants requires little attention and it is possible for one man to discharge from the store a measured quantity of ice.

Table 9 Labour requirements (man hours/24 h)

t	10	20	50	100	200
Automatic					
Manager Clerk/Tel./Rec. Engine Room/Maintenance Harvesting Storing Discharge	8 8 - - 24	8 8 24 - - 24	8 8 32 - - 32	8 8 32 - - 32	8 8 32 - - 48
Total	40	64	80	80	96
Block					
Manager Clerk/Tel./Rec. Engine Room/Maitenance	8 24	8 8 32	8 8 48	8 12 56	8 12 56
Harvesting Storing Discharge	24	48	72	96	112
Total	56	96	136	172	188

Table 9 compares the labour requirements for different sizes of automatic and block type plants, but it must be stressed that in both cases the labour requirement will be dependent on design. No inclusion is made for trimming of vessels at the quayside and the block plant is assumed to be of the continuous harvest type. Twenty four hours availability of ice is assumed.

Water

The basic water requirement for ice making will be slightly more than the equivalent quantity of ice allowing for losses and bleeding to drain to prevent the build up of salts in circulated water systems. Additional water will be required for domestic purposes and for condenser cooling. The quantity required for condenser cooling depends on the amount of heat to be rejected, the temperature of cooling water and whether the system recycles the water or not. The requirements of a shell and tube condenser in tons of water per ton of ice produced is given in Table 10.

Table 10 Requirements of cooling water for a shell and tube condenser

Water temperature (°C)	Tons of water/t of ice
10	15
15	25
20	40
25	60
30	125

An evaporative condenser will use between 0.25 and 0.5 t/ t of ice depending on design. Precise requirements can be calculated from manufacturer's figures making allowance for windage, evaporation and bleeding. Shell and tube condensers employing a water tower and recirculation will have similar requirements to the evaporative condenser.

Water used for ice making should meet microbiological standards of potable water and be free of excessive solids or salts, which may detract from the physical properties of the ice produced. Problems may also be experienced when using pure water for flake ice production as the ice produced adheres to the drum and is difficult to remove. This problem may be overcome by the addition of salt to the make-up water.

Maintenance

Although maintenance costs for financial planning are often budgeted as a fixed percentage of the capital cost the actual costs are obviously dependent on the type of ice plant, its delivery system, the quality of servicing it receives and the age of the plant. Maintenancecosts are often higher in the very early life of a plant and in its later years.

Between 2 and 5 percent of the capital cost is commonly allocated per annum but actual costs can vary greatly year to year. Block plants generally have higher maintenance costs due largely to repair or replacement of moulds which may have an average life of only a few years. The cost of replacing the moulds of a 100-t/d plant would be in the order of US$ 60 000 to 100 000. Block plants also require more attention to the building structure due to evaporation of the brine and its effect on the steelwork although this can be reduced by the use of inhibitors.

ICE STORAGE

The type and size of an ice store will depend upon the type of ice, the demand for ice and the patterns of demands. It is common for ice stores to be 2-4 times the daily production. Physical dimensions of stores of given capacities are included under the description of each type of store and are based on manufacturers published information. As a comparison Table 11 lists the volume and area requirements of different types of stores with the forms of ice commonly stored in them.

Table 11 Volume requirements of ice stores

Type of store	Type of ice	Volume (m) of store for given capacity				
		50 t	100 t	200 t	500 t	
Silo a/	Flake	460	800	1 600	4 000	
Rectangular bin	Flake	196	356	656	1 608	
Rectangular bin	Plate	163	297	547	1 340	double bin
Rectangular bin	Tube	292	412	735	1 685	
Block store	Block	150	300	600	1 500	

Table 12 compares the area requirements of the types of store with different forms of ice commonly stored in them.

Table 12 Area requirements of ice stores

Type of store	Type of ice	Area (m) of store for given capacity			
		50 t	100 t	200 t	500 t
Silo a/	Flake	42	62	124	310
Rectangular bin	Flake	41	75	138	337
Rectangular bin	Plate	34	62	115	281
Rectangular bin	Tube	35	55	98	211
Block store	Block	36	75	150	375

a/ Inclusive of the air jacket space surrounding the silo

Block Ice

Block ice is usually stored in block form and crushed on demand. It is stored at temperatures between -4°C and -2°C and can be stacked one block upon another. Wooden battens can be used to separate blocks to prevent fusion but this is not normally a problem. Care should be taken when stacking tapered blocks which should be stacked to a maximum height of 2.2 m depending on block size or crushing of the lower blocks may result with danger to workers. The smaller sizes of block (e.g., 25 kg) may be stacked manually but larger sizes normally require some kind of lift or inclined conveyor to raise blocks one above another to a height at which they can be dragged using tongs. It is normal to base capacity upon a storage volume of 2-3 m^2/t of ice stored.

Flake Ice

Flake ice is usually stored in silo or bin with the ice-maker located over the store The silo system usually requires slightly less ground area for a given storage capacity but will usually be taller than a bin system and will have a greater capital cost due to the civil work involved, particularly if the site has an exposed location where windage is an important design factor. To prevent bridging of the ice and to assist the flow of ice on discharge, an agitator rotates within the silo driven by an electric motor from above. The silo works on the first in-first out principle of storage but ice build-up can occur on the walls and requires occasional removal either by blows to the outside of the silo or by removal from the inside manually.

Normally a trap door is provided in the top of the silo for access. The trap door is connected to a safety mechanism that isolates the agitator mechanism. The capital cost of the smaller capacities (40 t and below) are expensive compared with simple bin systems. For capacities much in excess of 100 t it is usual to install multiple units.

Many patented systems of handling and discharge exist but the mechanical rake and end gate system is common. The system operates on the first in-last out principle and requires emptying occasionally to remove stale and packed ice build-up at the bottom. To help overcome this problem the double bin system was developed which employs two bins and which enables one bin to be emptied while the other is filled.

It also discharges at a height convenient for loading but requires nearly twice the ground area of the rectangular bin and for this reason is rarely used for stores of large capacity. Other bin systems exist that operate on a first in-first out principle by removal of ice at the bottom of the store but the systems are expensive compared to the rake and end-gate systems and are more commonly used on large industrial application.

Discharge of flake ice is usually by screw and/or belt conveyor. Pneumatic discharge systems need very careful design for application with flake ice or a pulverized soft and wet product will result

Plate Ice

Plate ice is usually stored in a bin system similar to that described above for flake. The storage capacities which are for flake ice will be increased slightly for plate due to the higher packing density of plate compared to flake ice. A factor of 1.2 should be applied to the stated capacities when used for plate ice. Plate ice is commonly discharged by screw and pneumatic systems.

Tube Ice

Tube ice is commonly stored in bin systems using a scraping bucket or grab device to handle the ice in the store. The operation is controlled from an observation platform with a view through a window into the store. The walls and sometimes the floor are usually protected by timber planking to prevent damage by the bucket.

The observation window is also suitably protected by a guard. The bucket or grab transfers the ice via a screw discharge conveyor to an automatic weighing device which can be read from the observation and control platform. Usually only small packaged units of up to 15 t/d (production) are located above the store and it is usual that the ice is delivered from the ice-maker at ground level to the store by vertical and horizontal screw conveyors. The ice is distributed within the store by a centrifugal or gated belt device.

Simple Bin Stores Using Manual Discharge

Simple bin stores of capacity up to 50 t are available or can be made, without expensive weighing or automatic discharge systems, suitable for manual discharge.

CAPACITY OF SILO

	TONS	10	20	30	40	50	60	80	100
A	mm	1075	1445	1755	1870	1985	2135	2355	2660
B	mm	3350	4140	4120	4530	5140	5550	6365	6480
C	mm	2080	2475	2760	3080	3190	3340	3555	3860
D	mm	2870	3830	4465	5015	5260	5580	6060	6700
J	mm	1500	1550	1550	1550	1600	1600	1700	1700
K		depends on type of ice machine							
L	mm	4000	5000	5700	6200	6500	6800	7300	7900
M	mm	1000	1000	1700	1700	1700	1700	1700	1700
N	mm	8000	9100	9400	10100	10900	11400	12500	12950
LB	mm	3475	4150	4150	4500	4800	5175	5175	5175
LS	mm	2000	2000	2500	2500	2500	2500	3000	3000
Q	Kcal/h	4000	5500	7000	8000	9000	9500	11300	13000

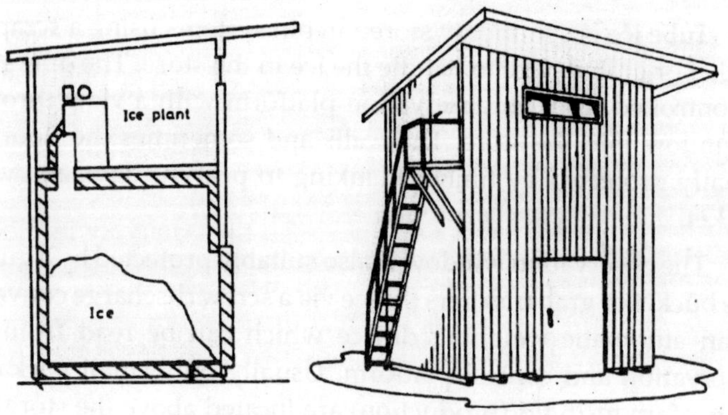

Figure 17. Simple bin store

Other bin stores are available that use modified insulated transport containers that require a minimum of site preparation and civil works. The stores can be purchased with ice-maker as a packaged unit. The plant can be considered as semi-permanent and should need arise it is possible to relocate it.

Ice Store Costs

The capital cost of ice stores will depend upon the type and capacity of the store, location and the specification or otherwise of associated equipment for refrigeration, handling, discharge and weighing, etc. The costs given for ice plants include stores but as a guide the following costs given in Table 13 apply to the store only. The costs are inclusive of steelwork, cladding, refrigeration plant (silo), piling and civil works, insulation, erection, automatic discharge (where stated) and weighing (where stated).

Table 13 Ice store costs

Type	Cost/t of store capacity (US$)
Silo store with automatic discharge and weighing equipment (50 t plus)	2 000-3 000
Bin store with automatic discharge and weighing equipment (50 t plus)	1 400-2000
Refrigerated block store	700-1 000
Bin store with manual discharge	500-1 000

Ice Crushers

Mechanical ice crushers are available suitable for crushing block ice to a desired particle size within given limits. The loss on crushing is minimal (a fractional percentage), but loss on delivery will depend upon the distance involved and methods of delivery and may amount to more. If the factory is located some distance from the quay or from the point of usage then delivery in block form from the plant to a remote crusher may reduce loss during transit as loss is dependent on a surface heat transfer effect.

Crushing by hand, by ice pick and/or mallet is a laborious process that often leads to losses as high as 15-20 percent and larger particle sizes which are less efficient for the rapid cooling of fish.

Table 14 gives details of capacity, size of blocks, particle size, installed power and costs for electrically driven mechanical crushers.

Table 14 Ice crushers

Capacity (t/h)	Block size - cross section	Particle size (cm)	Inst. power (kW)	Cost ex works (US$)
2	25 x 11.5	3.75-7.5	1	3 500
6	25 x 20.3	1.9-6.5	3	5 000
10	50.8 x	3.0-6.5	7	7 300
40	20.3	3.75-7.5	10	15 300
90	81 x 30.5	3.75-7.5	20	20 500
	167.5 x			
	30.5			

ICING OF FISH

In order to maintain optimum fish quality, fish, once caught should be chilled to 0ýÿC as quickly as possible. Ice chills fish by surface heat transfer either by direct contact between fish and ice or by cold melt water running over the fish surface. Therefore the more ice in contact with the fish the quicker the cooling rate. Even quicker cooling can be achieved by immersing the fish in a mixture of iced water which allows maximum surface heat transfer.

At Sea

Bulk stowage

Bulk stowage of fish is the stowage of iced fish in pounds usually formed by fitting portable pound boards into fixed vertical stanchions in the fish room. First a bed of Ice is laid in the pound then a layer of fish, followed by further layers of ice and fish to a maximum depth of half a metre. A second platform is then formed of pound boards and the process repeated with generous applications of ice to the top, bottom and sides of each pound.

Pound boards are commonly made of aluminium, plastic or wood; the first two being preferable for reasons of hygiene and handling. Corrugated extrusions give strength to the boards and provide for better draining of contaminated melt water to the edges of the pounds.

As a guide to icing, an ice to fish ratio of 1:2 is commonly used in temperate climates for voyages up to 18 days and 1:1 for tropical climates. Ice requirements can be theoretically calculated from a knowledge of the thermal properties of the fishroom, the length of voyage and the ambient temperatures but in practice the requirements can only be realistically based on experience. At the end of the voyage at least a little ice should be remaining in each pound. It is better to use a little too much ice than not enough.

The stowage density of bulked fish is about 0.50 t/m^2 of fishroom.

The major disadvantage of bulked stowage is the problem it presents on unloading. Digging out of the fish by hook and shovel is a laborious, slow and difficult process that results in a greater degree of damage than that caused by unloading of boxed fish. It is also a common cause of industrial injury. The rate of unloading depends upon the method used, the vessel, quality and number of the work force and the size of the catch. As an example, 60 men (22 fishroom, 5 hatchway, 5 landing boards, 5 winch, 23 dragging baskets) can unload 125 t from a distant water vessel in 6 h using baskets on gin wires from a landing leg above the hatches. This does not include for setting-up or weighing, sorting or laying out on the market.

Boxed stowage

Boxed stowage is similar in principle to bulk stowage except that instead of building up pounds in the fishroom the fish is mixed with ice in boxes. Similar quantities of ice are used for boxing as used for bulking except when the box is to be used for onward road transportation on landing without additional icing. In such cases extra ice is required to suit the onward journey.

The stowage density of boxed and iced fish is in the order of 0.37 t/mýÿ depending on the size and design of the box. The advantages of boxed stowage are: easier and quicker unloading, reduced handling of the fish and reduction in the damage to fish. As an example of the unloading rate, 8 men (3 shore, 3 fishroom, 1 hatch, 1 winch) can unload 10 t of boxed fish from an inshore vessel, using a gin wire and hooks from the vessels derrick, in less than an hour.

Chilled sea water

Stowage of fish in sea water chilled by ice and held in insulated fixed tanks offers several advantages over other methods of stowage, particularly for pelagic species not normally gutted at sea but which are caught in bulk and require rapid chilling.

Stowage in an ice-water mix minimizes crushing which causes damage and weight loss and rapidly cools the fish by efficient heat transfer. By design of the deck pounds and scuttles or pumping system the fish can be rapidly transferred from the deck or net to the tanks with the minimum delay and disruption to fishing effort.

The amount of Ice required to cool a tank of fish and water and maintain it in a chilled condition will depend upon the size of the tank, the effectiveness of the insulation, the ambient temperatures and the length of the trip. Allowance should also be made for ice loss during the outward passage to the fishing grounds. As a guide to icing a water: Ice: fish ratio of 1: 1: 4 is commonly used for 3-4 days stowage in insulated fixed tanks in temperate climates and 1: 2: 6 in tropical climates. For a given tank installation and prevailing conditions the quantity of Ice can be estimated.

The calculation of the ice required to combat the heat gain

to a fixed tank system is more complex and depends upon the method of construction of the tanks and particularly the way in which the tank lining is supported and insulated. The total heat gain to the tank Is the sum of the heat gains through each tank surface and for each tank surface the gain must be calculated depending upon the conditions that exist on that surface. For example, if we consider an outer wing tank side, we must calculate for the transfer of heat through the tank surface and frames above the water line on the outer surface of the vessel, and the transfer through the surface and frames below the water line, as each surface area will be subject to different outer temperatures and will have different surface heat transfer coefficients. The overall heat transfer coefficient for a tank surface, or part of, is a function of the outside heat transfer coefficient, the inside heat transfer coefficient and the thickness and conductivity of the insulation, the tank lining and the outer steel surface, as given by:

$$\frac{1}{u} = \frac{1}{h_0} + \frac{x_1}{k_1} + \frac{x_2}{k_2} + \frac{x_3}{k_3} + \frac{1}{h_i}$$

where,

u= overall heat transfer coefficient (kcal/mýÿ h deg C)

h_0= outside heat transfer coefficient (Kcal/mýÿ h deg C)

X_1= thickness of plate, ship's side (m)

k_1= conductivity. of plate, ship's side (kcal m/mýÿ h deg C)

X_2= thickness of insulation (m)

k_2= conductivity of insulation (kcal m/mýÿ h deg C)

X_3= thickness of tank lining (m)

k_3= conductivity of tank lining (kcal m/mýÿ h deg C)

h_i= Inside heat transfer coefficient (kcal/mýÿ deg C)

For each surface then or part of surface the heat transfer can be calculated through the surface, and frames, from the above given the temperatures that exist on each surface. From the total heat transfer to the tank the quantity of ice needed to offset the gain can be calculated from the latent heat of the ice. Examples of surface heat transfer coefficients for a typical steel tank installation are given in Table 15.

Table 15 Surface heat transfer coefficients

Surface	Heat transfer coefficient (kcal/m h deg C)
Deckhead, moving air outside	29.3
Tank floor, fresh water tanks beneath, gently agitated water	515.9
Engine room bulkhead, still air on engine room side	7.1
Ship's side above water, moving air outside	29.3
Ship's side below water, moving air outside	1 719.6
Tank inside, gently agitated water	515.9

Tank stowed fish for human consumption is usually unloaded by brail or by pump. Unloading rates of up to twenty t/h can be achieved using a brail of one t capacity. Rates of 30 t/h or more are possible with pump discharge but with a higher degree of damage to the fish, particularly if the fish has been previously pumped from the net.

Ashore

Upon landing the fish should be maintained at or as near as possible to its chilled temperature of 0ýÿC through the market, processing and distribution chain.

If the fish is to be laid out on the market for display prior to sale additional ice or re-icing may be required to prevent the fish becoming warm and spoiling. This practice however is often ignored because of the problem it causes in viewing the fish on sale and checking the fish weight. This can be overcome by weighing the fish prior to icing if a market kit is used, or check weighing a small number of boxes if the vessels boxes are used for display. In any instance, the fish should be moved as quickly as possible and kept out of direct sunlight.

If the fish is to be re-iced at the market after sale for transport to the processing factory or secondary market the quantity of ice required will depend on the temperature of the fish, the length of the journey. The ambient temperature and the thermal protection given to the fish by the fish box and the transportation vehicle. The quantity of ice theoretically to chill the fish can be calculated from Table 16 which gives the weight of ice required to cool 1 kg of fish from selected temperatures. In practice more may be required.

Table 16 Quantity of ice theoretically required to cool 1 kg of fish to 0°C

Temperature (C)	kg of ice to cool 1 kg of fish to 0C
30	0.34
25	0.28
20	0.23
15	0.17
10	0.12
5	0.06

The quantity of ice required to maintain the fish in chill will depend on the ambient temperature, the insulative properties of the box and vehicle, and the position of the individual box within the load. A box in contact with the sides or base of the lorry will require more ice than a box in the centre of a stacked load which is protected from heat ingress by the boxes surrounding it, but it is not always practicable to ice boxes according to their later loading position within a lorry. In practice the amount of ice used on any given journey is based on experience but it can be estimated given the box and vehicle dimensions, thickness of insulation and thermal conductivities.

Typical conductivities of fish box materials are as follows:

Aluminium	172 kcal m/m h deg C
Wood (wet)	0.24 kcal m/m h deg C
Plastic	0.40 kcal m/m h deg C

Table 17 shows the quantity of ice required in practice to chill and maintain in chill boxed fish under selected surrounding temperatures, for both a single box and for a stack of 35 boxes.

Table 17 Ice requirements to chill and maintain in chill fish held in individual boxes and within a stack of boxes

	Melting of ice per box of 50 kg fish					
	1 box			**35 boxes**		
Surrounding temperature (C)	+30	+20	+10	+30	+20	+10
Fish chilling (kg)	21	14	7	21	14	7
Keeping (kg/h)	3	2	1	1	0.7	0.3

Not only is the quantity of ice of importance in a box, also the distribution of ice within a box.

If the fish is already chilled then the ice is required to combat the heat gain through the box and can be packed round the fish against the box surfaces, but if the ice has to chill the fish also then the ice should be well mixed with the fish.

If a box of fish was iced on the top only for example, even with the calculated and correct amount of ice for cooling and maintaining the fish in chilled condition. the fish would not be efficiently chilled even if the box was well insulated. This is because the cooling is largely achieved by surface heat transfer and if the bulk of the fish are not in contact with the ice, the transfer of heat through the bulk of the fish will require a considerably longer cooling period. (Some cooling may be achieved by the passage of melt water.) In effect the fish at the bottom of the box would be 'insulated' from the ice by the fish above them. For the same reason it takes a substantially longer time to cool a large fish of a given weight than to cool the same weight of smaller fish to a uniform chilled temperature.

Handling Ice

Manual operation

The simplest form of delivery and icing system is a manual operation. Ice can be removed from bins using hand rakes and shovels and carried to the point of use in tubs, carts or box pallets. Such handling is satisfactory for small operations. Icing at sea is still most commonly carried out with a shovel using an axe to break the surface crust. It is a laborious and physically exhausting job.

Conveyors

Straight line fixed conveyor systems are the most common means of handling ice within the ice factory and for delivery of the ice to the fishing vessel or road vehicle. Where the ice factory is sited a long distance from the quayside specially designed tipping trucks or tow units can be used to deliver ice to petrol-driven inclined conveyors that deliver the ice through the fishroom hatch to the hold. Conveyors are normally supplied with galvanized finish and can be Insulated. Simple screw conveyors can act as elevators up to a maximum angle of approximately 300 to the horizontal depending on the type of ice although other types of screw conveyor will deliver vertically. Dished belt conveyors will handle ice horizontally and slated belt conveyors at inclinations up to approximately 400 or more depending on the design.

Screw conveyors should not be left with ice in them when stopped but should be run for a short period (depending on the length of the conveyor) after the supply of ice to them has been stopped to ensure that all the ice is discharged. Failure to do so can result in jamming the conveyor and burning out the drive motor. It is possible to fit an auto-delay to the conveyor motor to ensure this.

Pneumatics

Pneumatic conveying of ice is often used to deliver ice from the ice factory to the fishing vessel or road carrier but can also be used for Icing of fish at sea. Pneumatic conveying on shore is more viable at greater distances (over 30 m), and when multiple stations are to be served. Delivery rates of 10-60 t/h are obtainable.

The system must be carefully designed for the handling of ice, particularly flake ice which can be reduced to a slush by impact at high velocity on the duct walls. To reduce the heating effect of the blown air it is usual to precool the air through a heat exchanger. Bends should be a minimum radius of 1.5 m.

At sea a pneumatic delivery system with flexible discharge hose greatly simplifies the task of icing fish in the hold. The hose has only to be directed at the fish and velocity of discharge throws the ice to the required box or pounds eliminating the laborious task of shovelling ice. In a large fishroom hold numerous coupling points can be provided for connexion of the discharge hose. Figure 18 shows a pneumatic distribution system for a distant water wet fish trawler fitted with its own ice plant.

CHILL STORES

Design and Construction

Chill stores suitable for holding fresh fish should operate in the range 0-4°C in order that the fish, which should always be stored in ice, is not frozen or partially frozen. In this range the fish is efficiently cooled by ice melt water and thereby kept moist. Each store should be specifically designed or selected for the operating conditions.

It is not the intention here to describe the detail design of a store but to discuss the operating conditions will effect the design and costs.

In the design of the store the refrigeration engineer will require to know the total weight of product, the turnover of product, the handling methods, the temperature of the fish going into store, ambient temperature, water cooling temperature (if water cooled) and The cost/availability of electricity, water, labour and space. The refrigeration duty will be based upon:

- the heat entering through the insulated structure
- the heat from the incoming fish
- the heat from electrical equipment (including lights)
- the heat from men working in the store
- the heat from air exchange

Methods of estimating these loads are available in refrigeration manuals.

Physical Parameters

For a given weight of fish the physical size of the store will depend greatly upon the handling and stowage methods and the requirement for access and withdrawal of stock. For small stores manual loading and unloading is quite feasible and often the cheapest solution but may not be economic for larger stores where large volumes of fish need to be moved quickly. Mechanical handling offers advantages in quicker and cheaper movement of stock and in some cases lower capital and running costs. Fork lift trucks for example enable higher stacking of fish in the chill room and it is therefore possible to design the room to take advantage of this. Consider for example the two following rooms of the same internal volume.

The first store has a total surface area (including floor) of 192 mýÿ and the second only 160 mýÿ. Now the heat gain through the surfaces of the store is proportional to the surface area so the running cost of the first will be higher than that of the second. Additionally, the capital costs of construction will also be higher for the first store as the higher surface area will require extra insulation and floor surfacing. Considering also two stores of the same form but of different sizes as shown below:

The volume per unit surface area in the first case is 0.33 and in the second case 0.66 which illustrates the economy of scale with increasing size. The cost per unit volume, both capital and running, will be lower for larger stores. An important consideration in the sizing of a store is the requirement for access and withdrawal of stock. Consider the stowage of small pelagic species stored in ice at a fish: ice ratio of 3:1 in a box of dimensions 600 x 368 x 214 mm. Each box will contain approximately 25 kg of fish which represents a stowage volume of about 2 m°/t with no allowance for internal clearance, access or handling and evaporator. In practice a minimum of 3 m°/t should be allowed for storage where internal access is not important. This might be the case, for example, where fish is to be unloaded from a fishing vessel one afternoon, held overnight and brought out for sale on the market the following morning. It would not matter that the first fish into the store was the last to be brought out (FILO).

Where access is required for selective withdrawal of stock, possibly of different species or ages, the stowage volume required will depend on the degree of selection required and the methods of handling. Between 4 and 8 m°/t is commonly used and provides sufficient access for most situations. Figure 19 shows the floor plan of a store designed to hold 5 t of boxed fish, handled manually, based on a design figure of 4 m°/t. The plan assumes a box of dimensions 600 x 368 x 214 m loaded with 25 kg of iced fish. The boxes are stacked nine high (1.93 m). The internal dimensions of the store are 3.50 x 2.50 x 2.30 m (high).

Mechanical handling of stock in a chill store is unlikely to reduce the required floor plan area or volume required per t of fish if access is required for selective withdrawal of stock, as the aisles required for access and manoeuvring of a standard fork lift truck will be far In excess of that required for manual handling, and will not be compensated by the extra height of stacking. If selective withdrawal is not necessary then mechanical handling will enable a smaller ground area particularly for larger stores (over 30 t).

The width required for a 90ýÿ stacking aisle for a standard fork lift (1-1 1/2 t capacity) will vary depending on manufacturer and pallet size, but for a 1 200 x 1 200 mm pallet it is approximately 3.25 m. Special aisle trucks are available, designed for warehouse operation that will turn in a much narrower aisle width but are seldom used in the fish processing industry, Probably because the standard fork lift is better suited to other duties that it may have to perform in addition to work in the chill store.The plan assumes a stack-only box of dimensions 600 x 368 x 214 mm using a fish: ice ratio of 3:1 and holding 25 kg of fish. The boxes are shown stacked to a height of nine boxes. The aisles (1.2 m) between boxes provide access for selective withdrawal from any box column. The internal volume per t of fish stored is 7.0 m^2/t.

The plan assumes a pallet size of 1 200 x 1 200 m with 24 boxes per pallet (6 boxes per layer x 4 layers). The pallets are shown stacked three high. The layout allows access to any pallet stack. The floor plan area is less than that required for manual handling but the volume. required per t of fish is greater at 8.0 m^2/t. The pallets are stacked four high and it is assumed that they each hold 450 kg of iced fish. Access is available to any box pallet stack. The internal volume required per t of fish stored is 9 m°/t.

In each of the foregoing examples the capacity is greatly increased if selective with drawal is not required.

The examples assume the storage of small pelagic species in a stack only box at a fish: ice ratio of 3: 1. Different fish species, box sizes and designs and fish: ice ratios will effect the space requirement for storage. For larger species such as cod, the volume should be increased by as much as 16 percent depending on size of fish. If a stack-nest type of box is used the volume should be increased by as much as 18 percent depending on size and design. At a 2:1 fish: ice ratio it should be Increased by 20 percent, and at 4:1 decreased by 12 percent.

Technical Parameters

As previously stated, it is not the intention here to describe the detail calculation of technical parameters such as optimum insulation thickness, refrigeration duty, etc., and selection of equipment which is the province of the refrigeration engineer and must be determined for each case in point.

However, it is often necessary for purposes of feasibility study or planning to have some estimation of the consumption of electrical and water services and the size and connexion of installed electrical machinery. To this end the following three cases serve as examples which by interpolation might, for the purposes stated, be used in the planning of chill stores.

Three sizes of store are considered: 5 t, 25 t and 100 t (assuming a design factor of 4 m^2/t) under increasingly adverse (higher) conditions of ambience, water temperature and product temperature.

In each case a daily product loading of half the capacity of store is assumed. The 5-t store operates under an ambient air temperature of 20°C with mains water supply (for cooling) at 15°C and with incoming fish at a temperature of 5°C. The 25-t store is assumed to operate under an ambient air temperature of 30 ýÿC with mains water supply at 25 °C and with incoming fish at a temperature of 10°C.

The 100-t store is assumed to operate at an ambient air temperature of 40°C with mains water supply at 30 °C and with incoming fish at a temperature of 20°C. The refrigeration duty is respectively: 2 654 kcal/h, 11 335 kcal/h and 75 113 kcal/h. Note that the difference in power consumption between air cooled and water cooled in Cases I and II is small, but in Case III it is significant (103.5 cf. 79 kW).

The installed power rating of the compressor for Cases 1 and II is the same, but for Case III is only 55 kW for water cooled compared with 90 kW for air cooled. For Case III it is assumed that the motor is star delta connected, whereas for Cases I and II the motors might be connected direct on line. There is a slight difference in the compressor running times: Case I - 18 h; Case II - 19 h; and, Case III - 20 h.

Costs

The capital costs of the three chill stores considered in Section 7.3 and specified in Table 18 are shown in Table 19. The total capital cost is given for stores with air- and water-cooled condensers, and broken down into component costs of the store and fittings (insulation, surface finishes, doors and floor), refrigeration equipment (air and water cooled) and erection (including electrical wiring).

Analysis of the costings illustrates the economy of scale mentioned in Section 7.2 with the cost per unit volume of store dropping from US$ 600/mýÿ for Case I, the 5-t store, to less than US$ 300/mýÿ for Case III, the 100-t store (which operates under higher ambient conditions).

For the smaller stores the difference in capital costs between the air and water-cooled plants is not significant and might well depend on the manufacturer's product range as to which works out the cheaper. In Case III (100 t) the water-cooled plant works out to be nearly US$ 6 000 cheaper (5 percent less than air cooled), but again could be influenced by the manufacturer's product range or actual site conditions.

The question of choice between an air-cooled or water-cooled condenser would normally be made on running costs depending on the relative costs (or possibly availability) of electricity and water. Note the difference in power consumption given in Table 18 between air-cooled and water-cooled plants and the water consumption in the case of the water-cooled condenser.

Table 18 Wet fish chill rooms - specification and utilities consumptions

	Case I	Case II	Case III
Storage capacity	5 090 (5t)	25 454 kg (25 t)	101 818 kg (100 t)
Approx. room sizes (internal)	2.44 W x 3.66 D x 2.28 m H (8 ft W x 12 ft D x 7 ft 6 in H)	4.487 W x 6.09 D x 3.5 m H (16 ft W x 20 ft D x 11 ft 6 in H	6.09 W x 15.54 D x 4.50 m H (20 ft W x 51 ft D x 14 ft 9 in H
Room volume	20.39 m (720 ft)	104.2 m (3 680 ft)	426 m (15 045 ft)
Thickness and type of insulation walls and ceiling	75 mm (3 in) foamed polyurethane	100 mm (4 in) foamed polyurethane	125 mm (5 in) foamed polyurethane
Floor insulation	75 mm (3 in) foamed polyurethane	100 mm (4 in) foamed polyurethane	125 mm (5 in) foamed polyurethane
Thickness and type granolithic cement on floor	50 mm (2 in)	75 mm (3 in)	100 mm (4 in)
Means of handling	Manual	Manual	Fork lift
Ambient air temperature	20°C (68°F)	30°C (86°F)	40°C (104°F)
Mains water temperature	15°C (59°F)	25°C (77°F)	30°C (86°F)
Condensing temperature: Air cooled Water cooled	32.2°C (90°F) 30°C (86°F)	43.3°C (110°F) 45°C (113°F)	54.4°C (130°F) 41.1°C (116°F)
Fish entering temperature	5°C (41°F)	10°C (50°F)	20°C (68°F)
Design room temperature	0-3°C (32-37°F)	0-3°C (32-37°F)	0-3°C (32-37°F)
Maximum daily product load to be cooled in 24 h	2 545 kg (5 599 1b)	12 727 kg (27 999 1b)	50 909 kg (111 999 1b)
Number and size of door	one - 0.762 W x 1.92 m H (2 ft 6 in W x 6 ft 3 in H)	one - 0.90 W x 1.92 m H (2 ft 11 in W x 6 ft 3 in H)	one - 2 W x 3 m H (6 ft 6 in W x 9 ft 10 in H)
Refrigerant	R12	R12	R12 Air; R22 Water
Number of air coolers	one	Two	Four (electrical defrost alternated)

approx. compressor speed (rpm)	1 450	1 450	1 540 air 1 050 water
Refrigeration duty Compressor (kW): Air cooled (kW) Water cooled (kW)	2 654 kcal/h (10 500 Btu/h) 1.1 (1 1/2 hp) D.O.L. 1.1 (1 1/2 hp) D.O.L.	11 335 kcal/h (45 000 Btu/h) 5.5 (7 1/2 hp) D.O.L. 5.5 (7 1/2 hp) D.O.L	75 113 kcal/h (298 200 Btu/h) 90 (125 hp) S.D. 55 (75 hp) S.D.
Approx. compressor running time	18 h	19 h	20 h
Total power absorbed: Air cooled (kW) Water cooled (kW) Water consumption (shell and tube condenser)	5.1 (6.84 hp) 4.85 (6.5 hp) 772 litres/h (170 g.p.h.)	11.9 (15.95 hp) 11.44 (15.34 hp) 4 654 litres/h (1 025 g.p.h.)	103.5 (138.8 hp) 70 (93.87 hp) 23 154 litres/h (5 100 g.p.h.)

Table 19 Chill store costs (US$)

Case	Cold store and fittings	Refrigeration equipment		Erection inc. electrical	Total capital cost	
		Air cooled	Water cooled		Air cooled	Water cooled
I	4 200	6 100	6 920	1 500	11 800	12 620
II	9 120	17 820	18 300	4 200	31 140	31 620
III	34 600	72 200	64 400	10 300	117 100	111 300

FISH BOXES

Material

Fish boxes are commonly made of wood, aluminium or plastic. Wood was the traditional material for fish boxes but is being increasingly replaced now by plastic for reasons of cost and problems of usage.

Wood is extremely difficult to clean properly, has a short life compared to plastic and requires frequent repair. Aluminium is easier to keep clean but is prone to damage by rough handling making them difficult to stack. Aluminium boxes are also very noisy to handle on a concrete market surface. Plastic (thermoplastic high density polyethylene) is easy to clean, will withstand rough handling and allows detail design requirements to be incorporated in the mould.

Strength

Boxes used in the fishing industry are often subjected to abuse by rough handling and should be designed to withstand such treatment. At sea they are often subjected to fierce gravitational forces caused by vessel motion. Particular attention should be given to the design of the handles or recesses for lifting, which should be strong enough to carry the load of three full boxes as vessels often discharge this many boxes using hooks to the handles of the bottom box of three. A stack formed from loaded boxes should be rigid and not tend to flex.

Weight

The tare weight of plastic and aluminium boxes is constant and not subject to age or environmental conditions. Wooden boxes will absorb water and can almost double in weight when saturated making accurate weighing of fish in the wooden boxes difficult and increasing transport costs.

Drainage

Fish boxes should be designed to have good drainage in order that the melt water, which contains bacteria from the fish, is allowed to drain away. Some designs enable the melt water to drain through the box so that melt water from one box will then pass through the box beneath it thereby providing the maximum cooling effect from the cold water. The disadvantage of this design is said to be cross contamination by bacteria carried by the water from box to box. Other designs prevent this by drainage to the outside of the box but with the loss of the cooling effect of the water.

Nesting

The ability of a box to nest when not in use is very useful when storing empty boxes, particularly for applications within fish markets and box pools. Nesting boxes will give a reduction in storage volume of up to 60 percent depending on design compared to stack-only types. The nesting type of box does however suffer the following drawbacks, particularly for use at sea:

i. nesting boxes can only be safely stacked one directly above another which results in a poor utilization of fish room space on fishing vessels which have curved sides;

ii. the tapered sides required for nesting result in a poorer volumetric efficiency (capacity) compared to a stack only box;

iii. the boxes are more difficult to stack and unstack manually as they require lifting out or into position. Care also has to be taken to ensure the box is right way round or damage to fish by crushing will result and the stack becomes unstable;

iv. the reduced base area of a nesting box produces a stack that is less stable than stack only boxes with interlocking grips.

An advantage of some nesting designs is the ability to lift boxes by the lips by a modified fork lift or pallet truck which eliminates the need of pallets.

Stacking

Boxes that stacking only should incorporate interlocking grips between boxes to prevent sideways movement of boxes in the fish room at sea or during transportation on shore as shown in Figure 24. Plastic and aluminium boxes can be designed with this facility but not wooden boxes. The grips should allow staggered patterns of stacking for better utilization of fish room space. Stack only boxes are volumetrically more efficient than nesting boxes but require far greater storage volume when empty.

Useful Life

The life of any box is plainly dependent on factors of design, material, construction, degree of use and methods of handling but as a comparison aluminium and plastic boxes might last 6-10 years, respectively, whereas a wooden box might last only 2 years and require repair during that period. In the calculation of requirements of boxes over a period, pilfering and loss should be accounted for, and in this respect, plastic boxes offer an advantage over wooden or aluminium inasmuch as the material is of little use for alternative purposes.

Dimensions and Capacities

The dimensions and capacity of a fish box should reflect factors of fish size, packing density of fish and ice, use of mechanical handling equipment and whether a one-or two-man lift is desirable. It may also need to be compatible with existing boxes, possibly wooden, and of a capacity that is a traditional unit of sale. The length should be such that fish will not overhang the boxes or have to be forced into it and the depth should be such as to allow for fish and ice without crushing. If the box is to be used on a roller conveyor care must be taken to ensure a continuous bearing surface along the base of the box in the direction of intended motion. The volumetric capacities and dimensions of different types of boxes are given in Table 20.

Table 20 Dimensions, capacities and costs of fish boxes

Type	Material	Volume (1)	Cost a/ (US$)	External Dimensions (mm)	Tare weight (kg)	Capacity of fish (kg) d/
Stack nest	Plastic	30	9.80	800 x 450 x 150	2.5	17
	Plastic	60	11.50	800 x 450 x 270	3.7	35
	Plastic	100	17.70	900 x 495 x 355	5.0	56
Stack	Plastic	42	6.90	600 x 368 x 214	2.6	25
	Plastic	60	12.00	813 x 480 x 178	3.5	35
	Plastic	70	12.00	844 x 514 x 190	5.0	40
	Plastic	90	18.00	850 x 515 x 260	6.0	50
	Aluminium	76	29.80	832 x 370 x 260	6.0	43
	Wood	60	5.00	812 x 470 x 178	7.0 b/	35
Stack c/	Plastic	25	6.90	560 x 415 x 150	1.6	20 fillets
	Plastic	42	7.40	650 x 400 x 200	2.3	30 fillets

a/ Individual box cost based on an order of 1 000 boxes ex-works (1979)

b/ Dry weight which can almost double when water-logged

c/ Fillet boxes for use in the processing factory

d/ Approximate capacity of fish assuming a 2:1 fish: ice ratio (crushed block ice)

Fish Box Washing

It is essential that after use fish boxes should be thoroughly cleaned of all dirt, fish slime and scales. The method of cleaning to be employed will depend on the number of boxes involved and whether or not the debris has been allowed to dry onto the box.

Hose wash

With small numbers of boxes, a simple cold water wash may be all that Is required but if the debris has been allowed to dry on the boxes then an overnight soak In a detergent solution with scrubbing and hosing the following day will be necessary. Such treatment is not suitable for wooden boxes which cannot be satisfactorily cleaned.

High pressure cleaning

Portable high pressure cleaners can be used effectively to clean plastic and aluminium boxes but should not be used on wooden boxes.

The units are available with cold or hot systems operating at pressures. typically in the order of 70 kg/cmýÿ. Water consumption is about 12 litres/min and fuel consumption (hot water models) between 7-11 litres/h depending on temperature. Installed power is typically 2-3 kW. Approximate cost of cold water unit is US$ 1 500 and for a hot water unit US$ 3 500 (1979).

Box washing machines

Box washing by machine is by far the most efficient and quickest method of handl[illegible], large numbers of boxes with capacities in the range of 800-1 200 boxes/h.

Boxes are normally fed through the machine hung on a monorail conveyor or upside down on a roller or mesh belt conveyor.

The machines are typically made up [illegible]ree sections having a cold or warm water pre-rinse followed by a hot water detergent wash and finally a cold water rinse. A pre-wash is necessary to remove salt from the boxes as it reduces the effectiveness of the detergent. The washes are performed under pressures of 4.6-5.6 kg/cmýÿ if the debris is dried on, or slightly less if not.

Fresh water consumption for a unit of 1 000 boxes/h capacity would be about 5 t/h. This can be reduced by using pumped sea water for the pre and post-wash rinses but is not recommended because of possible contamination.

Power consumption will depend on throughput and water pressures involved, but is typically in the range 15-30 kW for a 1 000 box/h unit. Boiler oil consumption for heating water will be about 40 litres/h for a throughput of 1 000 boxes/h with a wash temperature of 70°C in a temperate climate.

A mildly alkaline detergent should be used for wood, aluminium or plastic boxes at a concentration of 1-2 percent solution.

For the 1 000 box/h machine a minimum labour force of four men will be required; two feeding the machine and two removing boxes plus additional labour depending on distances involved fetching and stacking boxes.

Price for complete unit, 1 000 boxes/h, including boiler and installation would be in the order of US$ 50 000-60 000 ex-works. The space required for the machine is not large (10 m long x 3 m wide x 2 m high), but relatively large areas must be allowed for handling of the boxes which will be dependent upon the type of box and maximum height, Stack/nest boxes in this situation have a big advantage.

INTERNAL TRANSPORTATION

The following range of machines or devices are commonly used as handling aids in the fishing industry. The range is wide and the use of a particular type will depend upon the total load to be carried, the unit load, cost of labour/machine, the distances involved, the floor surface and consideration of maintenance, etc. A brief description of types is given followed by Table 21 which gives prices and capacities.

Sack or Case Truck

A very simple manually operated two-wheel unit suitable for handling individual or small numbers of stacked fish boxes. Of limited capacity and normally used only over short distances.

Jak-tug

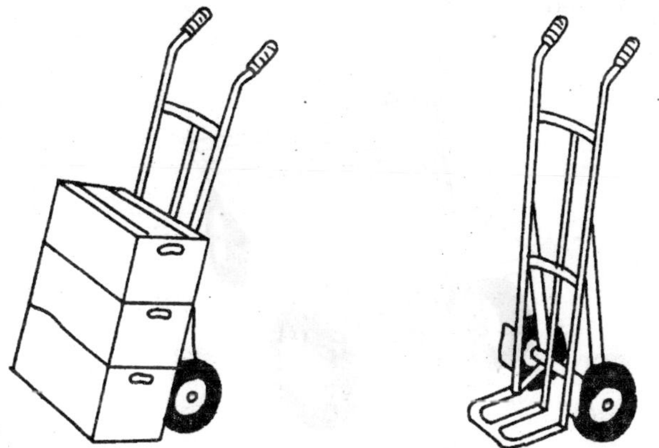

Figure: Sack or case truck

A rigid platform with wheels at one end which can be manouvered manually by the use of a towing unit that lifts and supports the other end. Restricted to use on more or less level floor.

Figure: Jak-tug

Hand-drawn Truck

A four-wheeled manually operated truck suitable for loads of up to 1 t. Steering is either by means of a turntable assembly or by a central pivot connected to the steering wheels by track rods. The track rod type of truck is preferred as the wheels remain under and give support to their respective corners on turning. The turntable type has a tendency to tip when negotiating sharp turns if unevenly loaded.

Figure: Hand-drawn truck

Electric Tow and Platform Trucks

They are basically battery-operated versions of the manual equivalent, which are better suited for handling greater loads over longer distances. Diesel versions are also available but should not be used in markets or processing factories where fumes and oil spillage can lead to contamination of product or be a hazard to operators. Gradients of 1:10 can be negotiated under load on floors that need not be smooth. Facilities are required for re-charging of batteries.

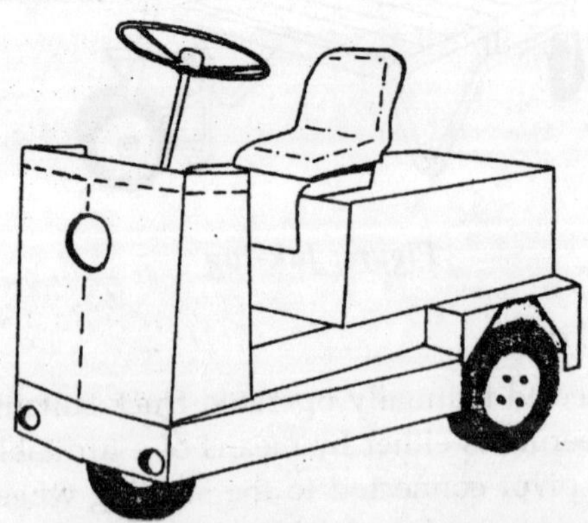

Figure: Electric tow tractor unit

Hand Pallet Truck

A manually operated truck for use with palletized loads. It consists of two forks and a steering unit. The forks are entered into the pallet which is raised by a hand-operated hydraulic pump, until the load is clear of the ground. Because of the relatively small wheels required for entry into the pallet the floor needs to be smooth and fairly level. Simple modification by means of a frame fitted over the forks makes it suitable for transport of stack/nest types of fish box. It is particularly useful in box, pool or factory operations and eliminates wear caused by dragging or pushing boxes over concrete surfaces.

Figure. Hand pallet truck

Powered Pallet Truck

An electrically powered truck similar to the hand pallet truck which can be pedestrian or rider-operated depending on design. Quicker in operation than the hand pallet truck and more suited to greater loads and distances. inclines. Can be operated on fairly uneven surfaces and inclines.

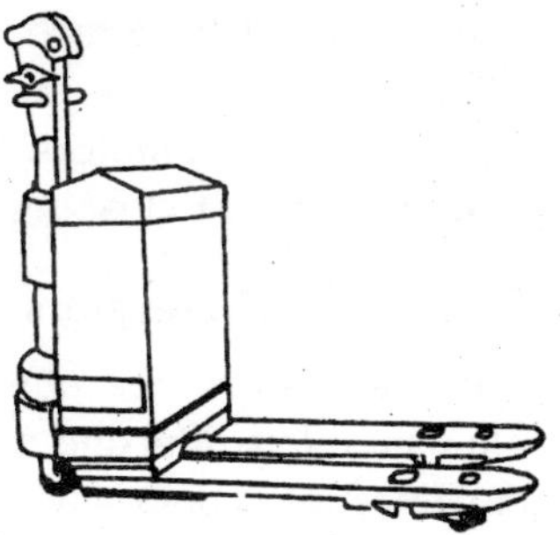

Figure. Powered pallet truck

Fork Lift Trucks

Similar to the pallet truck but with the capacity of lifting the load much higher. The forks do not contain the wheels and the load is balanced by the weight of the truck. They can be pedestrian or rider-operated depending on design, and are particularly useful for loading and unloading of road transport, stacking within chill stores and movement of processing machinery.

Figure. Rider electric fork truck

Roller Conveyors

Gravity roller conveyors are commonly used for the handling boxes or cases 'horizontally' over short distances. Powered rollers can be used over greater distances. Roller conveyors are available in unit lengths both straight or curved sections or can be specifically designed for intended application if the scale warrants it. Portable and expandable units are also available which can be removed or contracted to save space when not in use. Electrically powered units should be fitted with waterproof motors. Both powered and gravity rollers should allow efficient cleaning.

Powered Belt Conveyors s, and Elevators

Belt conveyors and elevators are commonly used for transportation of whole fish, fillets offal and other fish products. Wherever the product comes into contact with the belt then the belt used must be of food quality grade and should not contain metal fasteners but be of the continuous type. Smooth finish PVC type belts are restricted to horizontal applications for the transportation of fish but rough top belts can be used up to a maximum inclination of 150 to horizontal if dry or 200 if wet for most species of fish.

Table: In plant transportation capacities/prices

Type 1	Characteristics	Capacity a/ (kg)	Price b/ (US$)
Case truck	Can be made to suit box	200	70
Jak-tug	Platform 1 200 x 800 mm	500	140
Hand-drawn truck	Platform 2 000 x 1 000 mm	1 000	360
Electric platform truck	Platform 2 000 x 1 500 mm 2 kW motor 24 V battery 292 Ah at 5 h rate, 7.0 kWh Battery charger	Carry 1 000 and tow 1 000	5 500 750
Electric tow truck	1 kW motor 24 V battery 235 Ah at 5 h rate, 5.6 kWh Battery charger	Tow 3 000	3 500 700
Pallet truck	Width over fork 500-700 mm Fork length 800-1 200 mm Lifting height from 80-200 mm	2 000 (1 000-2 500)	400
Pallet truck (powered)	As for pallet truck Lift motor 0.35 kW drive 0.75 kW 16 V battery 216 Ah at 5 h rate, 3.5 kWh Battery charger	2 000 (1 000-2 000)	5 700 620
Pedestrian fork lift truck	Fork spread 260-800 mm Lift 3 300 mm Drive 1.0 kW Lift 2.0 kW 48 V battery 85 Ah at 5 h rate, 4.1 kWh Battery charger	1 000 (750-1 500)	10 500 680
Rider electric fork truck	Fork spread 260-900 mm Lift height 3 300 mm Drive 5.5 kW Lift 6.3 kW 72 V battery 235 Ah at 5 h rate, 16.9 kWh Battery charger		18 500 860
Roller conveyor (gravity)	35 mm roller; 180 mm pitch 600 mm wide point finish 3 000 mm length		100
Powered belt conveyor (food quality belt)	3 000 mm length 200 mm wide belt Drive 1.5 kW		1 600
Powered belt conveyor (standard belt)	3 000 mm length 600 mm wide Drive 2.0 kW		1800

a/ The capacity given is for the unit described under characteristics. The capacity range given below this is the range of units

b/ These are prices ex-works UK (1979)

The belt should be fitted with a brush and/or wire to clean the returning belt surface. Variable speed belts are often a great asset where manual operations are performed along the belt length allowing the speed to be varied to suit the number or capability of the operatives. Stop or emergency buttons should be clearly seen and easily reached. For inclinations greater than the above slated-belt elevators are commonly used for transport of fish and offal. In a continuous process the speed of the elevator and design and spacing of the slats should be such as to deliver the required volume of fish.

Flumes

Water flumes are often used for the transport of fish or removal of offal within a processing factory as an alternative to a conveyor and can be built in the case of offal removal beneath the floor provided easy access to the flume is allowed for. The degree of inclination and volume of water flow required is dependent on the fish size and type and the cross section and material of the flume. Trials have shown that for a 'U' section flume of 46 cm width, using 15 litres/min water flow, inclinations of between 5-10° to the horizontal were required depending on the size and type of fish.

PALLETS

Types

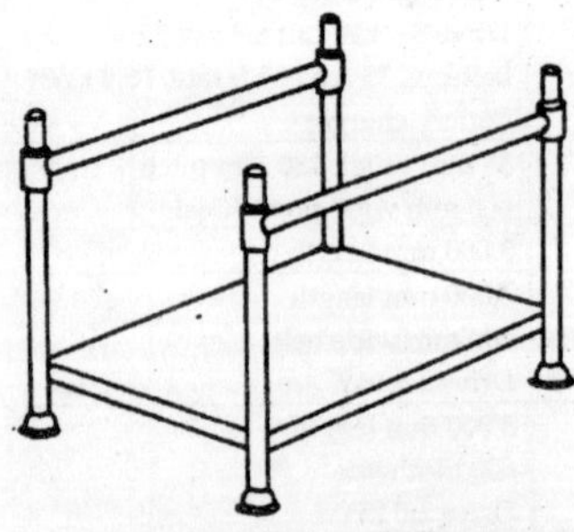

Figure: Post pallet

A pallet forms the base structure for a load in order that it may be handled by a fork lift or pallet truck. A unit load may be built up of a number of smaller unit loads, for example, fish boxes, on a pallet and transported and handled as a single load making the handling much simpler. They are particularly useful with regular shaped loads such as fish boxes when they can be stacked one upon another. If the loads are irregular, or if damage could occur due to crushing, post or box pallets can be used to stack one upon another as shown in Figures below.

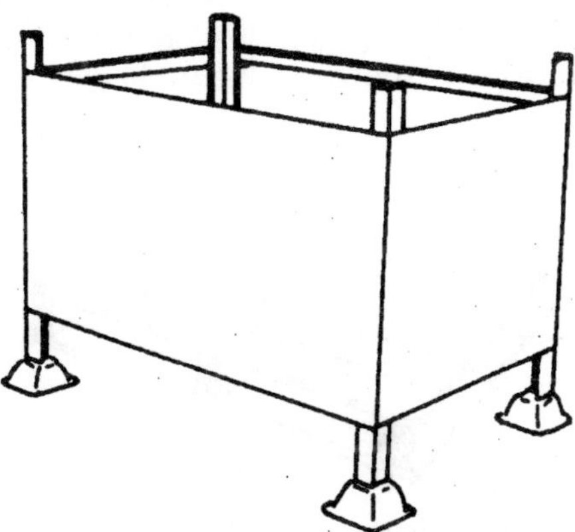

Figure: Box pallet

A pallet converter, is used with an ordinary flat pallet to achieve the same purpose as a post or box pallet.

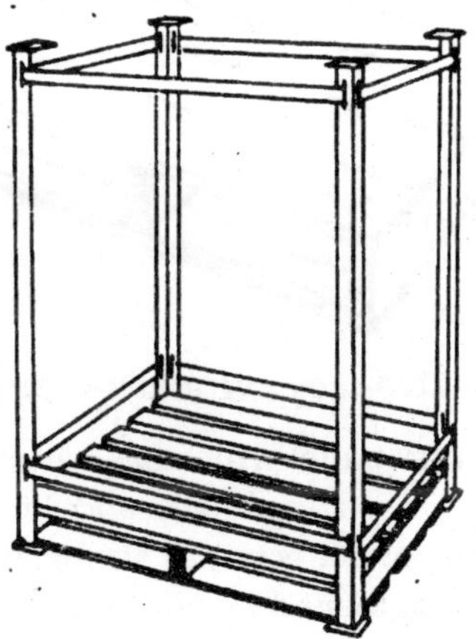

Figure: Pallet converter

The flat pallet is the most common type of pallet in general use in the fishing industry and can be of reversible or non-reversible type and of two-way entry or four-way entry. The reversible pallet shown in Figure 35 has identical top and bottom decks either of which can be load bearing and is of strong construction. It is not suitable for use with pallet trucks and is of two-way entry. The non-reversible pallet has only a load bearing top deck and is suitable for use with pallet and fork lift trucks. The two-way entry pallet consists of a top deck, three full length bearers and a suitably spaced bottom deck as shown in Figure below.

Figure: Two-way entry reversible

Figure: Two-way entry non-reversible

The four-way entry pallet which has advantages in handling consists of a top deck, 9 spacing blocks and 3 transverse stringers as shown in Figure below. It is not as strong as a two-way entry pallet but may be strengthened by additional slats to the base to form a perimeter base or cruciform base pallet as shown in Figure below.

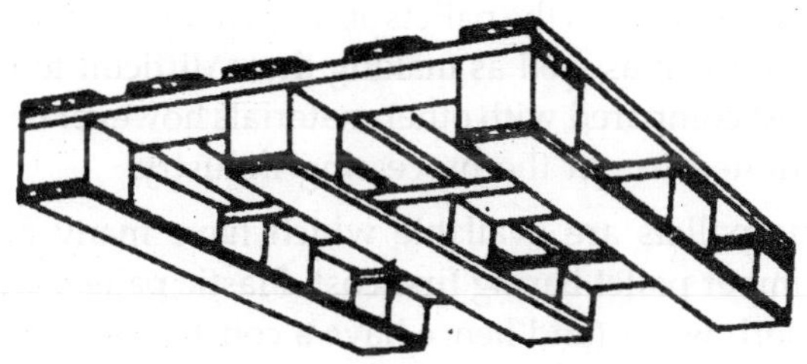

Figure. Four-way entry pallet

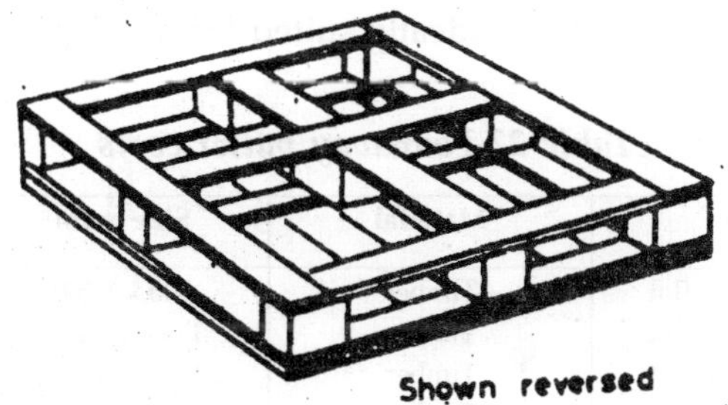

Figure. Four-way entry pallet with cruciform perimeter base

Winged pallets are pallets designed with decks that extend beyond the outer bearers for lifting by crane with spreader bar slings.

Wooden pallets may be specified in hardwood, which is better suited to rough handling or softwood and may have an unplaned or planed deck of open-boarded or close-boarded construction.

For use with pallet trucks the edge to the bottom deck may be chamfered to facilitate the passage of the load wheels of the truck.

Materials

Timber pallets may be of soft or hardwood but must be of good commercial quality that complies with international standards in respect to grain, knots, resin, splits and bark pockets. If screws are used in the manufacture they should not be hammered in and if nails are used they should be of the annular ring type.

As with fish boxes, timber is not well suited for applications in a wet environment as the pallets absorb water which can weaken and damage them as well as making them difficult to clean. The low first cost compared with other materials however has resulted in its continued use in the processing industry.

Plastic pallets are available which have many advantages over the timber pallet baring first cost. Plastic pallets are smooth. do not absorb water (and hence have a constant tare weight), are easy to clean and have longer life. Unfortunately the high first cost and the requirement frequently to leave the factory and be possibly lost has restricted its use.

Other than that it is ideally suited for use in the processing industry.

Table 22 Four-way pallet costs

Type	Material	Size (mm)	Cost (US$)
Non-reversible flat	Timber	800 x 1 200	8.80
Non-reversible flat	Timber	1 000 x 1 200	10.60
Non-reversible flat	Timber	1 200 x 1 200	12.00
Non-reversible flat	Timber	1 800 x 1 200	16.80
Reversible flat	Timber	800 x 1 200	10.80
Reversible flat	Timber	1 000 x 1 200	13.60
Reversible flat	Timber	1 200 x 1 200	15.60
Reversible flat	Timber	1 800 x 1 200	21.80
Non-reversible flat	H D polyethylene	1 000 x 1 200	30.00
Box with lid and drain a/	H D polyethylene	800 x 1 200 x 825	200.00
Box with lid and drain a/	H D polyethylene	930 x 1 250 x 720	280.00
Box with lid and drain a/	H D polyethylene	1 120 x 1 530 x 885	393.00
Box with sheet sides	Steel with painted finish	1 000 x 1 200 x 850	110.00
Box with sheet sides	Steel with galvanized finish	1 000 x 1 200 x 850	176.00
Post	Steel with painted finish	1 000 x 1 200 x 850	80.00
Post	Steel with galvanized finish	1 000 x 1 200 x 850	128.00

a/ Suitable for iced fish or ice chilled water and fish

Metal pallets are available suitable for use in a wet environment if properly protected, and designed so as not to damage tiled floors. Neither mechanical fasteners nor welding, fillets should project above the deck surface to an extent greater than the thickness of the deck materials. If the pallet is galvanized the surface should not be excessively rough.

Costs

The following table lists the ex-works cost of four-way entry pallets based on an order of 100 pallets. A slight reduction would be available in most cases for larger orders.

Life

The working life of a pallet depends entirely on its usage. the degree of rough handling it receives and its likelihood of loss. The life of a plastic pallet will be many time. that of a timber pallet justifying its high first cost if loss is not a serious problem.

Metal pallets will have a life between that of wood and plastic.

Plastic pallets can be distinctively coloured or permanently marked for identification to assist recovery and have little alternative use unlike the timber pallet that can be burnt for firewood or used for other purposes.

Specification

The International Organization for Standardization has recommended the following plan dimensions for standard pallets: 800 x 1 200, 1 000 x 1 200, 1 200 x 1 200, 1 200 x 1 600 (maritime use) and 1 200 x 1 800 mm. Each pallet can be specified for loading of 1 000, 1 500 or 2 000 kg.

The standard plan sizes are designed for loading of transport vehicles for industry in general and unfortunately are not always well suited for handling of fish in boxes. If standard flat pallets are to be used for handling fish in boxes consideration should be given in selection of pallet size to the stacking plan on the pallet for optimum utilization.

For example the 600 x 368 x 214 mm plastic box commonly used for small pelagic species fits well on the 1 000 x 1 200 (as shown in Figure 39), 1 200 x 1 200 and the 1 200 x 1 800 mm pallets.

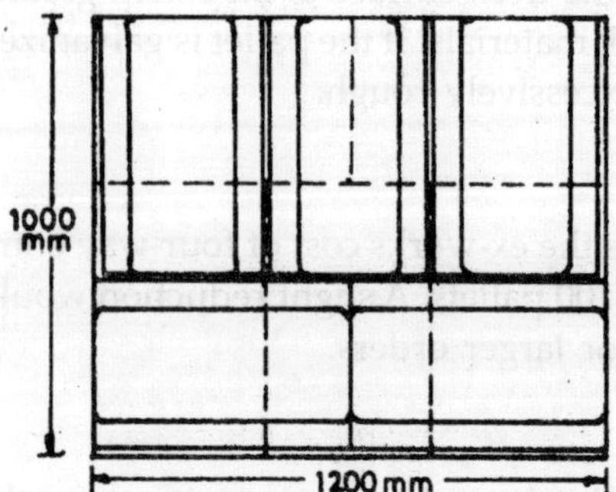

Figure. Pallet stacking plan for fish boxes

Some manufacturers of plastic boxes produce pallets and stillages not necessarily to ISO specification to suit their range of fish boxes.

Depending on the base design of the box subsequent layers of boxes may be repeated on the same plan or 'mirror imaged' as shown dotted in Figure above to give better stability.

For handling of fish in bulk particularly from the market box pallets often offer advantages in handling and storage.

INSULATED CONTAINERS

The use of insulated containers for stowage and transport of fish in iced sea water offers several advantages over traditional boxed and iced practises.

It is particularly suited to small pelagic species that are not normally gutted at sea but which are caught in bulk and require rapid chilling because of their high fat content.

Advantages:

1. Fish can be transferred rapidly from deck to the containers (by careful arrangement of the deck pounds and scuttles) with minimum delay and with minimum disruption to fishing effort.
2. Improvements in quality of the fish can be achieved due to the nature of stowage and to the reduction in handling of catch.

3. Rapid unloading of the vessels by removal of containers by crane and refitting with clean iced containers enables a quick turn about of the fishing vessel in port and substantially reduces the labour required for unloading.

Disadvantages:

1. The major disadvantage of the portable insulated container is the reduction of carrying capacity of both the fishing vessel and road transport.
2. For efficient operation of a container system the fishing vessel needs to be designed for the purpose. (Particularly scuttles. chutes. fishroom floors and unloading hatches.)
3. Fishing vessels are restricted to ports that can provide a change of containers.
4. A slight uptake of salt in fish flesh occurs with stowage in sea water.
5. Containerized catches are a less attractive proposition if the fish is to be sold through a traditional market of small unit sale and are better suited for sale direct to the processing factory.

Description

The following specification relates to a container developed for use in the UK for stowage and handling of herring and is shown in Figure 40. For similar use in tropical areas the insulation thickness should be increased to a minimum of 75 mm.

Container dimensions: 1 170 mm wide x 1320 mm deep x 1950 mm high

Insulation: 50 mm rigid foam polyurethane

Construction: Aluminium inner skin with GRP outer skin; hinged lid at top and drain at bottom

Capacity: 2.1 m

Empty weight: 175 kg

Aeration: By means of an aluminium lance. built into the container with coupling at top to manifold

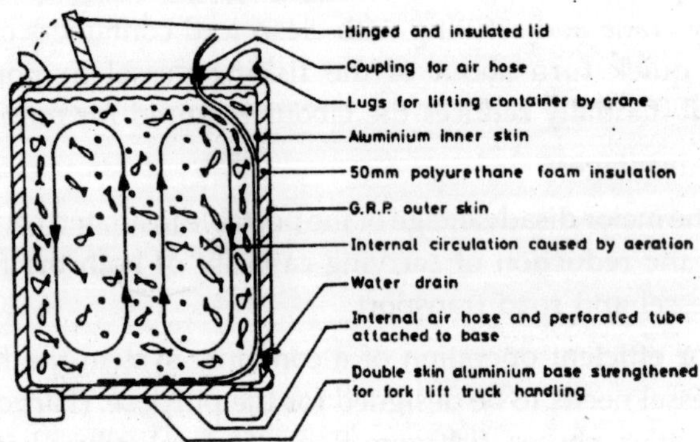

Figure. Chilled sea water container

Calculation of Fish, Ice and Water Quantities

For any given container.the ice:water:fish ratios can be calculated from a knowledge of ambient temperature. sea water temperature, the insulative properties of the container and the period of holding. The requirement of the ice is to cool the sea water and fish to 0°C, to counter the heat input from aeration and to counter the heat gain through the container. The requirement for water is to provide a medium that will enable rapid heat transfer from the fish and to hold the fish such that they are not subject to crushing. Table 16 (Section 6.2) shows the weight of ice required to cool, one kilogramme of fish or water to zero degrees centigrade. over a range of temperatures. It is assumed for the purposes of estimation that the specific heats are equal. Table 23 shows the weight of ice required to offset the heat gain through the container previously specified. It assumes a figure of 0.085 kg of ice loss per degree centigrade above zero per hour based on trials with the container. Table 23 Ice required for each day of stowage to counter heat gain through the walls of the container specified:

Ambient temperature (°C)	kg of ice/24h of stowage
35	72
25	51
15	31
5	10

Example Calculation

The following example calculation of ice:water:fish ratios assumes ambient temperature of 20°C. sea water temperature of 15°C. a holding period of three days and a container as previously specified.

Assume 1/5 volume of container for water

0.2 x 2.1 = 0.42 m.

Ice required to cool 0.42 m of sea water at 15°C from Table 16

=420 x 0.17 kg.71.4 kg

which occupies a volume of 0.08 mýÿ according to Table 3.

Ice required to counter heat gain through container for 3 days from Table 23 =123 kg which occupies a volume of 0.13 m.

For purposes of estimation the heat input by aeration will be assumed negligible.

Volume of remaining space within container

V_r = 2.1 - 0.42 - 0.08 - 0.13 = 1.47 m.

This volume will be the volume of fish plus the volume of ice required to cool the fish from 15°C to 0°C.

From Table 16 the ratio of fish:ice is approximately 1:0.17

volume of fish = 1.47 x 1/1.17 = 1.26 m. and

volume of ice = 1.47 x 0.17/1.17 = 0.21 m.

Total volume water = 0.42 m

Total volume ice = 0.08 + 0.13 + 0.21 = 0.42 m

Total volume fish = 1.26 m

From this follows:

Ratio fish:ice:water = 3:1:1 (by volume).

For higher conditions of ambience and sea water temperatures the above calculation can lead to unacceptably low carrying capacities and in these circumstances what is done in practice is to put in less ice and water initially than calculated but to later drain off some of the water and re-ice. It is important when doing this not to drain off all the water or the fish will no longer be floating in an ice water mix but subject to its dead weight which can lead to damage by crushing.

FISH WORKING PREMISES

Location

Perhaps the most important decision in the planning of a new processing plant or fish market is the selection of location. which is often critical to the success or otherwise of the venture. Once established it is most disruptive an costly to move site.

In evaluation of a site the following factors should be considered:

a. supply of raw materials (availability and distance from)
b. supply of labour
c. supply of services (electricity, water, sewerage)
d. disposal of effluents, offal, etc.
e. proximity to sales area or public
f. adverse factors effecting neighbours amenity (smell, noise, traffic)
g. civil costs of site development (clearing, levelling, etc.)
h. possibility of future expansion
i. access to good road/rail linkage or to vessel docking

Design

The initial cost of a building is only the first of a series of costs which will arise throughout its life. The building will require regular maintenance, heating and lighting, and cleaning and servicing rates and insurance will be incurred. Over the life of a building these running expenses will amount to two or three times its cost of construction and their incidence will depend on its planning and design.

The inclusion of running costs as well as initial costs within the costing concept considerably increases the value of cost analysis of the building design. There is often conflict between the effects of a design feature on capital and running costs, but the long-term nature of the investment necessitates a long-term evaluation of the effect of design on costs. As examples the amount of insulation specified will effect the heat gains or losses to or from a building, the use of natural lighting will influence power consumption and materials of construction will effect life, cost of insurance and the cost of maintenance and cleaning.

It is often helpful at the planning stage to draw up a process flow diagram showing the sequence of operations, proposed equipment, service requirements (water, electricity, steam) and the need for removal of wastes. This will assist in planning of layout and in assessment of total service requirements. Modelling techniques can also be usefully employed in planning layout.

The very diverse nature of fish processing operations, local practices and material availability make it impossible to describe within this text universal detail designs of fish working premises but the following notes relating to the principles of design, construction and materials are relevant.

Basic Shape

The basic shape or configuration of a building will be influenced by the nature of the process, the installed equipment and cost considerations. It may also be influenced by site restrictions or planning regulations. Most fish processing buildings and markets are single story or of a hybrid configuration with offices or stores above part of the processing area. In the design of a floor plan particular attention should be paid to product and materials handling; a through-flow design often being the simplest and most economical to operate.

Wholesale fish markets for example are often long narrow structures running the length of and a short distance from, the landing quay. Such a shape enables the maximum number of vessels to simultaneously unload at the market with the minimum handling of fish along the quay for market display. It also enables the maximum number of road vehicles access to the rear of the market for loading. The narrow width of building means reduced handling from vessel to road transport.

Floors

Floors should be hard-wearing, non-porous, washable, well-drained, non-slip and resistant to possible attack from brine, weak ammonia, fish oil and offal. Granolithic concrete is attacked by the continued action of strong brine and acids and accelerated by wear or damage of he surface. Nonetheless it is a popular and reasonably cheap surface for fish working premises. Clay tiles are more expensive than concrete but are harder wearing and less susceptible to attack by brine or acid. Care should be taken that they are layed level. Asphalt is not resistant to oils and is rather soft but is waterproof, fairly hard-wearing and less slippery when wet.

Areas subject to extra traffic may require metal tiles and the exposed edges of loading bays or steps may require some form of metal protection to the edge.

Fish working floor surfaces should be well drained and not allow the formation of water pools. A slope of 1:100 is usually sufficient for drainage but should not be greater than 1:40 which can be dangerous.

Drains

Where possible three drainage systems should be provided, one for domestic foul sewage, for storm water and surface drains, and the main drainage system for handling liquid wastes. Main drains should not connect directly to a sewer without an intermediate trap, and should be sufficiently barge to carry away all waste water without backing up or flooding. Floor drainage channels should have easily removable gratings which can be simply cleaned. It should not be possible for rodents to enter the building via the drainage system.

Walls, Ceilings and Fittings

The prime requirement of walls, ceilings and fittings in general is that they are easily cleanable. Walls should be smooth and waterproof and ideally surfaced with ceramic tiles to a height of at least 1 m. Pipe runs and electrical cables should be recessed into the wall or boxed in. If brine, salt or acids are used in processing special attention should be paid to protection of the steelwork or use of alternative materials.

Plant and Equipment

Again a prime requirement of plant and equipment is that it is readily accessible for easy cleaning. Preferably all plant should be raised off the floor on plinths. Electric motors and starters should be water (pressure) proof and good lighting should be provided where cutting, strapping or similar operations are undertaken. Copper and bronze are not suitable metals when in direct contact with food products; stainless steel or aluminium being preferable.

Area Requirements

The floor area requirements for fish handling operations depend on the throughput, the nature of the operation and methods of handling and storage. Within the building adequate space must be left for access and movement of fish, offal, ice, packaging materials, etc.

Area requirements for ice and chill stores are given in Sections 5 and 7 and fish box dimensions and capacities in Section 8. Careful consideration should also be given in the planning stage as to the requirements of access to the building and parking in its vicinity particularly in the case of fish markets.

It is typical of many small concerns operating on or near wholesale fish markets supplying a wet fish trade. It is assumed in this cage that centralised facilities of canteen, toilets, etc.,are available to staff.

The second example shown in Figure 42 is of a wholesale fish market designed for a maximum landing (mostly by vessel) of 200 t per day. The design assumes that a11 fish is laid out for display and that only one auction round is conducted each day. A boxed display is assumed with no stacking of boxes. The design makes allowance for storage of unloading and grading equipment, box cleaning and storage, and washroom facilities. Other facilities such as canteen, ice supply, banking and other offices are assumed to be available.

Costs

The capital cost of buildings suitable for fish processing or handling will vary enormously depending on location, design, materials of construction and size but as an indication of UK cost levels the following table gives the cost per square metre for three classes of building (excluding ground cost).

Table 24 Building costs

	US$/m
First class processing building including offices, storage, washrooms, etc.	250-350
Auction hall with washrooms and limited office facility	150-250
Sheds and open building of a simple nature	100-175

UNITS

The units of measurement used throughout the text are the conventional metric units, supplemented on a few occasions by the British units. However, for the convenience of the readers who may have familiar with the new International System of Units their conversion factors into the conventional units, and vice-versa, are given bellow for all the units that are not identical in the systems and that have been used in the text.

Selected Conventional Units

Power: 1 horsepower (hp) = 0.7457 kW

Specific energy: 1 kilocalorie per kilogramme (kcal/kg) = 4.187 kJ/kg

Heat flow rate: 1 kilocalorie per hour (kcal/h) = 1.163 W

1 British thermal unit per hour (Btu/h) = 0.2931 W

Specific heat capacity, mass basis: 1 kilocalorie per kilogramme, degree Celsius (kcal/kg deg C) = 4.187 J/g deg C

Thermal conductivity:1 kilocalorie metre per square metre hour degree Celsius (kcal m/mýÿ h deg C) = 1.163 W m/m deg C

1 kilocalorie per metre hour degree Celsius (kcal/m h deg C) = 1.163 W/m deg C

Thermal conductance: 1 kilocalorie per square metre hour degree Celsius (kcal/m h deg C) = 1.163 W/m deg C

Selected International System Units

Power: 1 kilowatt (kW) = 1.341 hp

Specific energy: 1 kilojoule per kilogramme (kJ/kg) = 0.239 kcal/kg

Heat flow rate: 1 watt (W) = 0.86 kcal/h

1 watt (W) = 3.412 Btu/h

Specific heat capacity, mass basis:1 joule per gramme degree celsius (J/g deg C) = 0.239 kcal/kg deg C

Thermal conductivity:1 watt metre per square metre degree celsius (W m/mýÿ deg C) = 0.86 kcal m/mýÿ h deg C

1 watt per metre degree celsius (W/m deg C) = 0.86 kcal/m h deg C

Thermal conductance:1 watt per square metre degree celsius (W/mýÿ deg C) = 0.86 kcal /mýÿ h deg C

Present and Future Markets for Fish and Fish Products

INTRODUCTION

At the twenty-sixth session of the FAO Committee on Fisheries, FAO was requested to identify how trade in fish and fish products could further benefit small-scale fisheries and generate additional income and employment within the sector. The FAO Committee on Fisheries emphasized that sustainable trade was dependent on sustainable fisheries management practices being in place.

Following the request, case studies were carried out in selected Latin American, African and Asian countries to study the importance of small-scale fisheries trade and identify opportunities for a better integration of small-scale fisheries into regional and international fish trade. The findings and recommendations of the case studies were presented and discussed at the tenth session of the FAO Sub-Committee on Fish Trade, held in Santiago de Compostela, Spain, 30 May to 2 June 2006.

While the definition of small-scale and artisanal fisheries varies between different countries and regions, the concept is commonly used to differentiate small-scale and artisanal fisheries from industrial fisheries. The FAO Working Group on Small-scale Fisheries characterized small-scale fisheries as "a dynamic and evolving sector employing labour intensive harvesting, processing and distribution technologies to exploit marine and inland water fishery resources". Of the 38 million people recorded by FAO globally as fishers and fish farmers, an estimated 69 percent are classified as small-scale. Women are involved in processing and marketing of small-scale and artisanal fish products. Their involvement in productive activities leads to increased household well-being because the income is spent on food and their children's education. Some women work for fish traders and others are employed as labourers in the fish processing industry. In India alone, it is estimated that around 700 000 women and children are employed in marine fisheries activities such as marketing, processing and net mending.

THE CONTRIBUTION AND ROLE OF SMALL-SCALE FISHERIES

The case studies presented in this Fisheries Circular highlight that small-scale fisheries are the backbone of the fisheries sectors in many developing countries. The lack of appreciation of the socio-economic importance of small-scale fisheries, however, often results in insufficient attention being given to the needs of the sector. The processing and trade of products from small-scale fisheries are important not only because these provide income and employment for many men and women in fishing and fish farming communities but also because the nutritional status of domestic and regional populations is dependent on these activities. Thus, supporting the needs of small-scale fisheries and ensuring their sustainability is a matter of survival for many countries.

Small-scale fisheries are important for the national economy. In Tanzania, for example, the 3 percent contribution of the fisheries sector to the country's GDP comes primarily from small-scale fisheries. As far as the contribution of small-scale fisheries to the marine fisheries production is concerned, significant differences can be observed between the Asian countries studied. While 70 percent of the Indian marine fish production can be attributed to the small-scale fisheries sector, the contribution of this sector to the marine fisheries production of Thailand and Malaysia has declined to 7.22 and 6.8 percent, respectively, because of the growth of the industrial fisheries sectors in these countries.

In the African countries studied, small-scale fisheries still account for the major part of the marine fisheries production, i.e. 95 percent in the case of Tanzania, 77-86 percent in the case of

Senegal, 80 percent in the case of Mozambique and 70-80 percent in the case of Ghana. In Ghana, the marine fish production by canoes in 2002 was many times higher than the production from purse seine and trawling vessels.

In the case of Brazil, the artisanal fisheries sector consisting of 326 696 fishers accounted for two-thirds of the marine fisheries production of 712 143 tonnes in 2003.

The inland capture fisheries sectors of most African countries are almost entirely composed of small-scale operators. In Tanzania, the freshwater fish production in 2004, entirely from small-scale fisheries, was six times higher than the marine catches while the value was four times higher. In the three Asian countries studied, the entire inland capture fisheries production comes from the small-scale fisheries sector.

In aquaculture, the contribution of the small-scale sector can likewise be considerable. In India, the entire production from inland aquaculture is contributed by small farms. Ninety percent of the shrimp farms are smaller than 2 ha and 5 percent are between 2 to 5 ha. Ninety-five percent of the production from coastal aquaculture comes from small farms. In Thailand, 82.5 percent of the inland fish farms are small, contributing nearly 70 percent to the total inland aquaculture production. Small farms account for 45 percent of the coastal aquaculture production.

India, Thailand and Malaysia have large domestic fish markets and fish products are sold in live, fresh/chilled and processed form. Cultured marine shrimp is the main export item, over 90 percent of which is exported in various product forms. The export earnings from shrimp in 2003 were US$847 million in India, US$160 million in Malaysia and US$1.2 billion in Thailand.

In India, Malaysia and Thailand, new opportunities lie in marketing more ready-to-eat products as there is an increasing trend in the consumption of seafood and an increased awareness that it is a healthy food. Products from artisanal fisheries can be marketed by giving value to the artisanal character of the product and the community, which produces it. Value-added products can also be marketed for different target groups in the domestic and regional markets. The use of different fishing gear can be an avenue for improving quality. An example is the use of longlines for catching tuna.

While the products from small-scale fisheries are presently still largely focused on the domestic market, in the African countries studied, regional trade is very important for supporting the protein requirements of poor people. In some places, the export of high value fresh fish to international markets can also be important. Small-scale fish processing and marketing are sources of employment and income for many rural households.

In the West African countries studied, a large proportion of fish from small-scale fisheries is consumed smoked and about 20 percent is consumed fresh, salted, sun-dried or fried. Cross-border trade of cured fish products is an important activity. In Tanzania, *dagaa,* the most important freshwater small pelagic species, is sun-dried after landing by women from fishing communities both for human consumption and for animal feed production. It is estimated that between 50 to 60 percent is used for animal feed production. Dried *dagaa* is marketed to neighbouring countries in the region. Nile perch, the dominant species in the export trade, is produced exclusively by small-scale fisheries and sold to processing factories. Nile perch rejected by processing factories is sun-dried, salted, smoked and fried for human consumption. The by-products are used for human consumption and also for fishmeal production. Dried trimmings, chests, maws and skins are Nile perch by-products, which are exported within the region and to Asia. Fillets are marketed to Europe, Australia, the United States of America, Asia and the Middle East.

In many African countries, women known as "fish mammies" dominate the fish trade and play the important role of providing informal credit. In Ghana, 90 percent of the artisanal fisheries production is handled by these fish mammies. In Mozambique, 20 percent of the 100 000 people in the artisanal fisheries sector are women.

Several opportunities exist for product diversification, value addition and improvement of product quality to develop and access new markets for small-scale and artisanal fish products. In Tanzania, the Fisheries Department Training Centre in Mwanza is experimenting with dried, spiced and smoked *dagaa* to improve shelf-life, quality and taste. In Latin America, between 70 to 80 percent of the fish produced by small-scale fisheries is marketed fresh while the rest is processed, most of it for domestic consumption. The phenomenon of increasing urbanization is concentrating domestic markets in big cities, where 80 percent of the 520 million people of the region are living.

The experience of the Women's Association of Betume in the northeastern state of Sergipe in Brazil shows that women can gain from increasing trade opportunities through value adding activities. The case studies from Mexico, Peru and Brazil suggest that the improvement of the quality of artisanal fish products is the key to a better integration of the sector into regional and international trade.

STRATEGY FOR A BETTER INTEGRATION OF SMALL-SCALE FISHERIES INTO TRADE

The first step to enable small-scale fisheries to benefit from fish trade should be the reduction of post-harvest losses and improving the quality of fish products for human consumption. This would increase the supply of fish without increasing fishing effort. Post-harvest losses owing to poor infrastructure, storage facilities and transportation as well as a lack of sufficient knowledge of proper and hygienic fish handling affect as much as 40 percent of fish landings.

While products for export are meeting high quality standards, those for domestic and regional markets are often processed through substandard hygienic methods and there is lack of consideration regarding the cleanliness of the drying, salting, smoking and processing environment as well as the holding, storage and distribution facilities. The potential exclusion of small-scale producers from international markets owing to the cost, difficulties and problems in complying with international standards such as HACCP and with standards imposed by supermarket chains also need to be addressed.

The case studies further suggest that efforts should be aimed at improving facilities for preserving fish onboard vessels, at working towards clean fish landing sites, increasing storage facilities, at the supply of ice as well as at improving roads that connect fishing communities to markets. Clean water is a basic necessity for ensuring hygienic conditions at fish landing and processing sites. Equally important is the improvement of technical support and extension services to enable fishing communities to access appropriate technologies, information and training opportunities. Small-scale fishers and processors need technical assistance and training on quality, proper fish handling procedures, storage, product diversification, value addition and packaging. Fishing communities should also be assisted in assessing their resources and identifying those that have potential for increased trade in domestic, regional and international markets.

The demand for small-scale fisheries products in domestic and regional markets can also be increased by offering these products closer to inland areas, increasing the number of retail outlets in urban centres and raising awareness regarding the health benefits of eating fish. Increasing urbanization and rising incomes provide new opportunities for ready-to-eat products from small-scale fisheries. Regional fish marketing information networks such as INFOFISH, INFOPESCA, INFOPÊCHE and INFOSA are important for providing information regarding demand, supply, consumption patterns, markets and prices. Thus, they should be supported and strengthened.

Small-scale fishers and processors can obtain better prices for their products by shortening the fish supply chain and increasing their bargaining and lobbying power. The formation of marketing cooperatives should be encouraged and existing associations of small-scale fishers and processors strengthened by providing support for institution building. Market surveys may be conducted and buyer-seller meetings and seminars may be organized regularly to promote trade opportunities and to help increase business contacts and information flow.

There is further a need to raise awareness among microfinance institutions regarding the needs of the small-scale fisheries sector for credit and savings services. At the same time, small-scale fishers should be informed about the existence of financial services and how these can be accessed. Such information is presently very limited or not available at all because of the geographical dispersion and remoteness of many small-scale fishing communities.

As emphasized by the twenty-sixth session of the FAO Committee on Fisheries, sustainable trade is dependent on sustainable fisheries management practices being in place. In this regard, the open access regime prevailing in many small-scale fisheries as well as the intrusion of industrial fishers into areas designated for use by small-scale fishers needs to be addressed by national governments. Experiences in many countries are demonstrating that co-management arrangements offer promising prospects for fisheries management. Following the review of the findings and recommendations of the case studies, the tenth session of the FAO Sub-Committee on Fish Trade held in Santiago de Compostela, Spain, 30 May to 2 June 2006 noted the importance of capacity building for small-scale fisheries and aquaculture including the necessity of organizing small-scale fishers and fish farmers, their participation in fisheries planning and management and of assisting the small-scale fisheries sector in product improvement and accessing markets.

The tenth session of the FAO Sub-Committee on Fish Trade further requested FAO to provide information on the contribution of small-scale fisheries to international trade including ornamental fish trade, lessons from pilot and case studies that may be useful to other countries and to elaborate on how post-harvest activities could be improved, taking note of environmental considerations. FAO was also requested to broaden the perspective and discussion on the topic to include, among other things, how developed countries could support the integration of small-scale fisheries into international trade through standards setting, intermediation, including financing issues, increasing the bargaining power of small-scale fishers in getting fair prices for their products, promoting traceability and ecolabelling of small-scale fisheries products and through value chain analysis.

PRESENT AND FUTURE MARKETS FOR FISH AND FISH PRODUCTS FROM SMALL-SCALE FISHERIES IN SOUTH AND SOUTHEAST ASIA – THE CASES OF INDIA, MALAYSIA AND THAILAND

Intergovernmental Organization for Marketing Information and Technical Advisory Services for Fish Products in the Asia Pacific Region (INFOFISH) Kuala Lumpur, Malaysia

The Fisheries Sector

The marine fisheries sector contributes 1.62 percent to the GDP of India, 1.5 percent to the GDP of Malaysia and 1.9 percent to the GDP of Thailand. The fisheries sectors of these three countries provide fish as a source of protein and employment opportunities for millions of people and are important earners of foreign exchange. In India, 1.284 million active fishers are involved in marine capture fisheries alone. In Malaysia and Thailand, 82 630 and 161 670 fishers, respectively, are involved in marine capture fisheries.

In the three countries, various types of fishing craft and gear ranging from traditional non-motorized boats to modern factory vessels are employed to harvest fishery resources. The number of fishing vessels in India is 280 491. Malaysia and Thailand have 30 751 and 118 436 fishing vessels, respectively. Most fishing craft range from 17 to 28 feet (5 to 8.5 m) in length and are powered by inboard or outboard engines as well as by sails in the case of traditional non-motorized boats. Various types of gears such as handlines, longlines, poles and lines, trawls, gillnets, purse seines and push nets are used. The majority of fishing vessels operate in coastal waters, which are fully and sometimes overexploited.

The definition of small-scale fisheries in the three countries shows much variation. Small-scale fisheries in Thailand are defined in terms of horse power, length of boat, labour employed and type of fishing gears used. Small-scale fisheries are defined as fisheries, which use traditional fishing gear and fishing boats with a maximum length of 10 m and a maximum engine horsepower of 30 hp, operate in the vicinity of the home base of the owner of the vessel and close to the coast. Small-scale fisheries in Thailand are further defined by low incomes, being a subsistence-level economic activity and by the fact that most of the labour is provided by family members and relatives. Small-scale coastal aquaculture also falls under the category of small-scale fisheries.

In Malaysia, the term small-scale fisheries is limited to fisheries using non-motorized and outboard powered fishing boats and to all inland capture fisheries. In India, more than two-thirds of the entire fishing and fish processing industries are classified as small-scale by the government.

SMALL-SCALE FISHERIES PRODUCTION

Capture fisheries production

The total marine fish landings in Malaysia increased by 0.86 percent from 1 272 million tonnes in 2002 to 1 283 million tonnes in 2003. As shown in Table 2.1, in India and Thailand, the total marine catch declined by 2.05 percent and 0.87 percent, respectively, over the same period. In all three countries, the bulk of the production came from inshore fisheries, which accounted for over 80 percent of the total marine fish production.

In Malaysia, traditional fishing gears operated by 23 565 fishing vessels produced 277 847 tonnes of fish valued at approximately US$0.25 billion while gears like hooks and lines produced 586 tonnes of fish valued at US$0.78 million. The contribution of small-scale fisheries to the marine capture fisheries production has been estimated at 6.8 percent in the case of Malaysia, 70 percent in the case of India and 7.2 percent in the case of Thailand. The contribution of small-scale fisheries to the marine capture fisheries production and its value is shown in Table 1.

Table 1: Marine capture fisheries landings in India, Malaysia and Thailand and the contribution of small-scale fisheries (SSF) in 2003

Country	Total landings (tonnes)	Share of SSF (%)	Quantity of SSF landings (tonnes)	Approximate value (US$ million)
India*	2 977 965	70.00	2 084 575	1 373.66
Malaysia**	1 283 256	6.80	87 261	75.93
Thailand***	2 620 582	7.22	189 387	229.78

Source:* Department of Animal Husbandry, Dairying and Fisheries, Government of India (2003) ** Department of Fisheries, Malaysia (2003) ***Department of Fisheries, Thailand (2003)

The inland capture fisheries production in India and Malaysia has shown an increasing trend during 2003 while the inland capture fisheries production marginally declined in Thailand. In India and Malaysia, the production showed an increase of 15 960 tonnes (2.51 percent) and 378 tonnes (10.96 percent), respectively, while in Thailand it decreased by 1 800 tonnes (0.91 percent). The entire production of freshwater fish from capture fisheries in the three countries is contributed by small-scale fisheries exploiting rivers, lakes and reservoirs as shown in Table 2.

Table 2: Inland capture fisheries production in India, Malaysia and Thailand and the contribution of small-scale fisheries (SSF) in 2003

Country	SSF landings (tonnes)	Share of SSF (%)	Approximate value (US$ million)
India*	651 719	100	292.90
Malaysia**	3 565	100	6.40
Thailand***	198 400	100	151.46

Source: * Department of Animal Husbandry, Dairying and Fisheries, Government of India (2003) ** Department of Fisheries, Malaysia (2003) ***Department of Fisheries, Thailand (2003)

Culture fisheries production

The aquaculture sectors in the countries studied have a large potential and can contribute significantly to their total fish requirements. All three countries have achieved progress in aquaculture production through careful planning and have adopted effective strategies for further growth. In all three countries studied, aquaculture production has shown an increasing trend for the last two years. It has been noted that the governments of these three countries, through their departments of fisheries, have laid emphasis on increasing output from aquaculture farms, irrespective of their sizes, whether small-scale or medium-scale.

While it is difficult to give an accurate figure of the number of aquaculture farms operating in the three countries, the number of farms in Thailand has been estimated at about 30 000. In India, the number of aquaculture farms is estimated at about 90 000 and in Malaysia the number of farms has been estimated at 10 000 to 15 000 farms. In all three countries, the aquaculture production for the year 2003 continued to consist mainly of crustaceans, molluscs and finfish varieties. The increase in production during the period 2002 to 2003 was 1.30 percent in India, 1.24 percent in Malaysia and 24.4 percent in Thailand.

Inland aquaculture

In 2003, inland freshwater aquaculture accounted for 94.9 percent of the total aquaculture production in India, 31.8 percent in Malaysia and 41 percent in Thailand.

In all three countries, the inland aquaculture production increased in 2003 compared with 2002. In India, it increased by 1.5 percent while the growth in Malaysia and Thailand was 12.4 percent and 9.7 percent, respectively. In India, the entire production from inland aquaculture comes from the small-scale fisheries sector. In Malaysia, the small-scale fisheries sector's contribution to inland aquaculture is estimated at 10 percent while in Thailand, nearly 82.5 percent of all freshwater aquaculture farms are small-scale farms, which account for about two-thirds of the entire freshwater aquaculture production. In all three countries, freshwater pond culture spearheaded production with species like common carp, catfish, snakehead, climbing perch and freshwater prawn contributing to the bulk of the production. Tilapia, freshwater eel and barbs are also produced by aquaculture in Malaysia and Thailand. The inland aquaculture production and the contribution of small-scale fisheries are shown in Table 3.

Table 3: Inland aquaculture production in India, Malaysia and Thailand and the contribution of small-scale fisheries (SSF) in 2003

Country	Total landings (tonnes)	Share of SSF (%)	SSF landings (tonnes)	Approximate value (US$ million)
India*	2 102 350	100.00	2 102 350	1 181.10
Malaysia**	53 211	10.00	5 321	7.17
Thailand***	361 125	68.60	247 732	217.79

Source: * Department of Animal Husbandry, Dairying and Fisheries, Government of India (2003)

** Department of Fisheries, Malaysia (2003) *** Department of Fisheries, Thailand (2003)

Coastal and brackishwater aquaculture

In 2003, coastal and brackishwater aquaculture accounted for 5.1 percent of the total aquaculture production in India while the contribution of this sector was as high as 68.2 percent in Malaysia and 59 percent in Thailand. In Thailand, the coastal and brackishwater aquaculture production increased in 2003 compared with 2002. In Malaysia, the coastal and brackishwater aquaculture production increased from 48 375 tonnes in 2002 to 49 544 tonnes in 2003 recording a growth of 37.3 percent. Over the same period, the Indian production declined by 1.5 percent.

In India, 90 percent of all brackishwater shrimp farms have a size of 2 ha and another 5 percent of farms have a size from 2 to 5 ha. Farms of a size of 5 ha or less are classified as small-scale. These farms account for 95 percent of India's coastal and brackishwater aquaculture production.

In Malaysia, only 5 percent of the coastal and brackishwater aquaculture production originates from small-scale farms while the contribution of small-scale farms to the coastal and aquaculture production in Thailand is estimated at 45 percent. In Malaysia and Thailand, coastal aquaculture and brackishwater aquaculture production focused on the farming of shrimp, marine finfish and crustaceans while in India, only giant brackishwater shrimp, i.e. black tiger shrimp and white shrimp were cultured. The production of coastal and brackishwater aquaculture in India, Malaysia and Thailand and the contribution of the small-scale sector are shown in Table 4.

Table 4: Coastal and brackishwater inland aquaculture production in India, Malaysia and Thailand and the contribution of small-scale fisheries (SSF) in 2003

Country	Total landings (tonnes)	Share of SSF (%)	SSF landings (tonnes)	Approximate value (US$ million)
India*	113 240	95.00	107 578	636.18
Malaysia**	113 949	5.00	5 698	8.45
Thailand***	454 275	45.00	204 423	983.65

Source: * Department of Animal Husbandry, Dairying and Fisheries, Government of India (2003)

Department of Fisheries, Malaysia (2003) Department of Fisheries, Thailand (2003)

Composition and value of small-scale fisheries production

In India, on the average, a tonne of marine fish produced by marine capture fisheries is valued at US$658.9 while a tonne of freshwater fish is valued at US$449.4. In Malaysia, the average value of marine fish is US$869.8 per tonne, while inland freshwater fish is valued at US$179.5 per tonne. In Thailand, the average value of marine fish is US$1213.3 per tonne and US$763.4 per tonne of freshwater fish. The small-scale fisheries production in India, Malaysia and Thailand, its source and its value are shown in Table 5.

Table 5: Small-scale fisheries production, value and sources in India, Malaysia and Thailand in 2003

Source/country		India*	Malaysia**	Thailand***
Marine capture fisheries	Q	2 084 575	87 261	189 387
	V	1 373.66	75.93	229.78
	UV	658.96	870.15	1213.28
Inland capture fisheries	Q	651 719	3 565	198 400
	V	292.90	6.4	151.46
	UV	449.43	1 795.23	763.41
Inland aquaculture	Q	2 102 350	5 321	247 732
	V	1 181.10	7.17	217.79
	UV	561.79	1 347.5	879.14
Coastal and brackishwater aquaculture	Q	107 578	5 698	204 423
	V	636.18	8.45	983.65
	UV	5 913.66	1 482.98	4 811.84
Total	Q	4 946 222	101 845	839 942
	V	3 483.84	97.95	1 582.68
	UV	704.34	961.76	1 884.27

Legend: Q: quantity in tonnes; V: value in US$ million, UV: unit value per tonne

Source: * Department of Animal Husbandry, Dairying and Fisheries, Government of India (2003) ** Department of Fisheries, Malaysia (2003) *** Department of Fisheries, Thailand (2003)

In 2003, the total value of the production of the small-scale fisheries sector of India was US$3 483.84 million while it was US$97.95 million in Malaysia and US$1 582.68 million in Thailand.

The major varieties of fish landed by various sub-sectors of small-scale fisheries are shown in Table 6.

SMALL-SCALE FISHERIES PRODUCTS AND THEIR MARKETS

Exports from small-scale fisheries show a similar product and market trend in all three countries studied. Products made from cultured marine shrimp were the main export items. Over 90 percent of cultured shrimp is exported in various product forms. In 2003, the export earnings from shrimp products in the three countries studied were US$846.69 million in the case of India, US$159.76 million in the case of Malaysia and US$1 274.76 million in the case of Thailand.

Marine finfishes were also exported in various forms. In Malaysia and Thailand, live marine fishes, mainly grouper and seabass, were exported as well as a small quantity of cultured live tilapia from freshwater sources. Other varieties of fish like carp, striped catfish, gourami, snakehead and freshwater prawn were consumed in the domestic market while a small quantity of freshwater fish from Thailand was exported. Freshwater shrimp and some varieties of freshwater fish were also exported from India. Table 7 shows the fish product forms in small-scale fisheries in the three countries.

Domestic Markets

In all three countries studied, there are large domestic markets and major portions of fish and fish products produced by small-scale fisheries are absorbed by these domestic markets either in live, fresh/chilled or in processed forms.

In India, over 95 percent of the fresh fish production is consumed within a distance of 200 km from the fish landing sites and the rest is distributed to places beyond 200 km from the coast. The distribution and product flow to cities and urban centres located far away from landing centres is increasing. It has been reported that due to inadequate facilities for preservation of fish onboard vessels, relatively poor hygienic conditions prevailing at landing centres, lack of supply and poor quality of ice and inadequate transport facilities, in most instances the fish sold in the domestic market is generally of relatively poor quality.

Table 6: Major varieties of fish landed by various small-scale fisheries sub-sectors in India, Malaysia and Thailand

Source/species	Finfish	Shellfish	Molluscs & others	Bycatch/mixed catch
Marine capture fisheries	Skipjack	Shrimp, crab	Squid, cuttlefish,	Lizard fish,
	Yellowfin tuna		Cockles	Threadfin
	Sharks & rays		Mussels	Goat fishes
	Sardines			
	Mackerels			
	Croakers			
	Anchovies			
	Tilapia			
	Kingfish Bombay duck*			
	Grouper Seabass			
	Pomfret			
	Hilsa*			
	Ornamental fish			
Marine/brackish water	Seabass, snapper	Shrimp, crab	Mussels, cockles	
	Grouper			
Inland capture fisheries	Carp	Freshwater		Eel
	Tilapia Catfish			Snakehead* Catfish**
	Ornamental fish			Barb**
Inland aquaculture	Carp	Freshwater		
	Tilapia**			
	Catfish**			
	Ornamental fish			

In India only

In Malaysia and Thailand only

The entire inland capture fisheries production is consumed in fresh form with the exception of a negligible quantity, which is dried. About 60 percent of fish from riverine resources is consumed locally while the production from reservoirs is sold in urban markets. As the demand for freshwater fish is very high, fish caught in remote northeast Indian states is also transported to these urban markets.

Fresh fish plays an important role for Malaysian consumers. A national survey conducted during 1998-99 indicated that on the average, expenditure for fish per household ranged from US$19 to 20 per month. In Malaysia, most of the fish and fish products from small-scale fisheries are consumed domestically either in fresh/chilled or dried form. A niche market also exists for live products from inland and coastal aquaculture such as tilapia, grouper and shrimp. Product diversification and value addition are increasing in the domestic market sector. A large portion of processed fish products such as breaded and battered products are becoming popular in the domestic market. Consumers generally prefer fresh and chilled fish products to frozen products.

Table 7: Fish product forms in small-scale fisheries in India, Malaysia and Thailand

Source/country	India	Malaysia	Thailand
Marine capture fisheries	Fresh/chilled, frozen, dried and traditionally processed	Fresh/chilled, frozen, dried and traditionally processed	Fresh/chilled, frozen, dried and traditionally
Inland capture fisheries	Fresh/chilled, dried and traditionally processed	Fresh/chilled, dried traditionally	Fresh/chilled, dried and traditionally
Inland aquaculture	Fresh/chilled	Live, fresh/chilled	Live, fresh/chilled
Coastal aquaculture	Frozen, live,	Frozen, live, fresh/	Frozen, live, fresh/ chilled

In the Thai domestic market, fish is generally consumed in fresh form. Fish consumption in all coastal areas is high. In Thailand, most of the fish and fish products from small-scale fisheries are consumed domestically.

There is also a market for live products from inland and coastal aquaculture such as tilapia, seabass, grouper and shrimp. Side by side with modern processing facilities for export, there are many small-scale processing facilities in coastal towns, which cater to the domestic market. The most common methods used for processing are sun-drying, salting, pickling, smoking and fermentation.

Export Markets

In all three countries studied, products from small-scale fisheries are already being exported to neighbouring countries and regional markets, though in only small volumes.

The regional markets for Indian seafood products from small-scale fisheries during 2003 consisted mainly of Southeast Asia and the Middle East. Markets for Indian seafood in neighbouring countries are Sri Lanka, Maldives and Bangladesh. While traditional products like dried fish have been exported to Sri Lanka and Maldives, chilled freshwater fish has been exported to Bangladesh. Southeast Asian countries such as Thailand, Singapore and Malaysia import mostly frozen pelagic fishes and chilled demersal fishes from Indian small-scale fisheries.

In 2002, the major nearby markets for Malaysian seafood products from small-scale fisheries were Thailand and Singapore. Regional markets for seafood from Malaysia are Viet Nam; China; China, Hong Kong Special Administrative Region (China, Hong Kong SAR); Bangladesh; Brunei; Cambodia; Macau; Maldives; Myanmar; Pakistan; Philippines; Sri Lanka and Taiwan Province of China (Taiwan PC).

In 2003, the major regional and nearby markets for Thai products were China, Hong Kong SAR, Malaysia, South Korea, Singapore and Taiwan PC.

MARKETING CHANNELS

Marketing channels for small-scale fisheries products varied significantly in the three countries studied.

India

In India, the marketing channels for marketing fresh fish from marine sources involve several intermediaries with different functions such as auctioneers, commission agents, wholesalers, retailers and fish vendors. Some intermediaries undertake multiple functions. The number of intermediaries and channels of distribution vary from region to region, state to state and district to district within a state. Depending on the distance of the markets and type of consumers, i.e. individual or institutional, the number of intermediaries increases or decreases. Fresh fish from marine sources often passes through one to eight hands before reaching consumers located in rural, urban and metropolitan centres. The bulk of the fresh fish, i.e. 70 percent passes through only one to three hands. In the case of institutional customers such as restaurants and hotels, not more than three intermediaries are involved. While the bulk of sales are carried out through auctioneers, about 40 percent of fresh fish is sold through commission agents or wholesalers.

As a result of the involvement of intermediaries in marketing, the net share of fishermen of the prices paid by the final consumers varies from 25 to 56 percent, depending on the states in which fish is traded.

The bulk of the dried fish passes through one to three intermediaries before it reaches the final consumer. The number of intermediaries varies from region to region. Processing of dried fish is done by fisherfolk (in 31.8 percent of the cases) as well as by traders (50 percent) and wholesalers (18.2 percent). Over 38 percent of dried fish is sold in other Indian states. Only 10 percent of dried fish is consumed in inland states without a coastline. Most of it is consumed in the northeast region of the country. Weekly markets in rural and suburban areas are the main markets for dried fish.

Almost the entire inland freshwater fish production with the exemption of a small quantity, which is dried, is consumed in fresh form. About 60 percent of the riverine fish production is consumed locally. Fish production from reservoirs is mostly sold in urban markets. Due to eating habits and the limited purchasing power of the people living close to fish landing sites, most fish is transported for sale to urban centres. Calcutta is the major urban consumption centre for freshwater fish. Fish from landing sites in all other states is sent to this city. As demand for freshwater fish is high in all urban centres in the northeastern Indian states such as Assam and other neighbouring states, fish caught in other states is also sent to these urban markets for sale.

The marketing channels and trade flows of freshwater fish are by and large similar across the country. Fishers sell their fish to wholesalers directly or through commission agents and retail intermediaries. Wholesale intermediaries sell fish to retail intermediaries either through commission agents or directly. Fish marketing through cooperatives and public fish marketing corporations is very limited. Even in the case where fishermen sell fish to cooperatives, private intermediaries are also involved in the marketing chain. In some cases, cooperatives become additional intermediaries in the marketing channel.

In the farmed fish sector, subsistence level production is consumed locally while commercial production is invariably sent to urban centres. The market intermediaries in commercial fish farming are limited to three or four. Fish farmers sell their fish to a fish trader, who transports the fish to urban markets, where it is disposed through auction to a wholesaler. The latter sells the fish he/she procures at auctions to a retailer. Thus, fish passes through three to four hands before it reaches the final consumer.

In the case of export of fish products, only a few intermediaries such as an auctioneer, a fish trader-cum-preprocessor and the final processor are involved in the marketing channel. In some fish marketing and export centres, fish is directly delivered to the processor-cum-exporter at a predetermined price. Farmed shrimp is either bought by the processor directly from the farmer or, if intermediaries are involved, their numbers do not exceed more than two or three. Intermediaries such as auctioneers and agents are also involved in certain states such as West Bengal, where auctioneer-cum-financier-cum-preprocessor and selling agents are involved in the marketing chain between farmers and processor-cum-exporters.

Malaysia

In Malaysia, three marketing levels can be distinguished, i.e. primary markets, intermediate markets and terminal markets, involving wholesalers at fish landing and production centres, wholesalers at consumption centres and retailers. The main activity at the primary markets is the assembling of catches and harvests for redistribution to other market levels with small portions of fish sold directly to local consumers.

Intermediate fish markets are involved in a number of overlapping fish trading activities. These markets operate not only as terminal wholesale markets to supply retailers but also as transit points for redirection of fish to other wholesale centres. Terminal markets are markets where fish is received from the wholesalers at fish landing sites and production centres. Here, the fish is redistributed through and to retail outlets, hotels, caterers and institutional customers as well as to final consumers.

The majority of fish catches are landed at privately owned fish landing complexes or jetties. The collection of fish and its distribution are efficiently organized by agents at landing centres, where fish is passed on to wholesalers or wholesalers themselves procure fish. Wholesalers at consumption centres play a major role in determining the price of fish. There is a strong and exclusive link between wholesalers at fish landing and production centres and wholesalers at consumption centres. Fish designated for wholesale markets in large consumption centres is collected by wholesalers at the production and landing centres and sent to the wholesalers at the terminal markets. Products from small-scale fisheries are directly purchased from fishers, vessel operators or farmers by wholesalers at the production centres, who transport the fish to terminal wholesalers, who in turn sell it to processors or directly to retailers. Figures 2.4 to 2.6 illustrate the marketing channels for different kinds of fish and fish products in Malaysia.

Thailand

The marketing channel for small-scale fisheries products in Thailand involves fishers, fish farmers, private commission agents, fish collectors, the Fish Marketing Organisation (FMO), commission agents, processing plants as well as wholesalers and retailers. Generally, fish landed at fishing ports and landing centres is channelled to local fish collectors acting on behalf of local processors and to private commission agents, who distribute their products to wholesalers and retailers in areas adjacent to landing sites and to agents or fish collectors, who arrange to transport fish to wholesale markets operated by the FMO. At these wholesale markets, most of the fish is sold by auctioning. The buyers at the wholesale markets are wholesalers, fish processors and retailers. The wholesalers sell the purchased fish to retailers at negotiated prices or sell fish in bulk to institutional consumers including supermarkets, restaurants and hotels on a contract basis. Figure 2.7 illustrates the marketing channels for small-scale marine fish in Thailand.

Traders involved in the marketing of the marine catch are fishers, fish agents, fish collectors, wholesalers and retailers. Fishers may be involved in fishing, selling and processing of fish at the same time. Fish agents receive fish from fishers and trade it by means of auction or negotiation. In remote areas, there are fish collectors, who buy from fishers or fish agents and sell to wholesalers in cities and also to retailers. Generally, fishers who land their catch at private fishing ports may have contracts with fish agents, who determine the price of fish.

In the case of cultured marine fishes, approximately 85 percent of the production is exported. The marketing system of cage cultured grouper is the same as that of live fish for export. Major intermediaries involved in marketing are fish brokers, collectors, wholesalers and exporters. The marketing process can be divided into two levels, i.e. the local level and the export level.

At the local level, the intermediaries involved are fish brokers and collectors. Most of them are large fish farmers. The roles of brokers and collectors differ with respect to marketing stages. Brokers are responsible for monitoring the movement of prices, informing farmers periodically, contacting fish collectors or wholesalers to sell fish, charging broker's fee from buyers and stand as guarantor for payments due from buyers while fish collectors are responsible for collecting fish from small-scale fish farms. In the case of export marketing, exporters function as agents and network with importing companies.

Fish farmers sell about 35 percent of their total production to fish collectors. Most small-scale fish farmers rely on fish collectors, who are experienced at selling fish and who have information and knowledge of fish market outlets. Fish that is not sold to collectors is sold at private fish assembly markets through private fish agents. After buying fish at farms, fish collectors also transport it to assembly markets for sale. Freshwater fish traded in assembly markets accounts for 72 to 75 percent of the total production. Fish agents, both FMO agents and private agents, as well as fish collectors at assembly markets distribute most of the fish to wholesalers, retailers and fish processors. Wholesalers distribute most fish directly to retailers while a small portion is sold to processors for export. Freshwater fish is exported mostly to a neighbouring country, i.e. Myanmar in fresh/chilled form. Retailers are the last link in the chain before the fish reaches consumers.

In Thailand, 93 percent of the total production of cultured freshwater fish is consumed domestically, 72 to 75 percent in fresh/live form and 18 to 19 percent in several processed forms. Export accounts for over 6 percent of the cultured freshwater fish production, mostly in fresh/chilled and frozen form. The marketing channel in rural areas involves small-scale fish wholesalers and retailers, who buy fish directly from local fish farms. Species of fish traded in rural areas are low priced species, which consumers can afford.

INVOLVEMENT OF WOMEN IN SMALL-SCALE FISHERIES

Women's involvement in fish harvesting and processing is similar in all three countries studied. Their participation in marine capture fisheries is mainly confined to shore-based activities such as fish marketing, fish handling, net making/mending and processing, i.e. sorting, grading, weighing, gutting and filleting of fish. These shore-based income generating activities undertaken by women are combined with their home making responsibilities. In general, women are at a disadvantage in comparison with men, as women tend to be given lower paid and unskilled jobs. In all three countries studied, women play various roles in fish marketing as agents, auctioneers, retail stall holders and fish mongers working individually, as a family unit or in rare cases as cooperatives.

India

Statistics on the employment of women in fisheries in India are lacking. As per the Indian Livestock Census of 1992, 533 800 women and children were employed in marine fisheries activities such as fish marketing, net mending and fish processing. It is estimated that the number of women involved in fisheries related activities has increased to 700 000 since the census was conducted.

Women are involved in all aspects of post-harvest fish handling, preservation, processing and domestic marketing. They also provide an integral link between producers and consumers in the domestic marketing sector.

In India, the fisheries related activities undertaken by women vary from state to state. In the states of Andhra Pradesh, Orissa and Tamil Nadu located on the east coast of India, fish drying/curing, fish marketing including auctioning and purchasing at landing centres, retail marketing of fresh and dried fish and net making and mending are the main areas of women's involvement. Compared with these three states, the participation of fisherwomen in fisheries activities in West Bengal, also located on the east coast, is rather limited. In the fishing villages of West Bengal, even fish drying and curing are done by men or women belonging to other communities.

In the states located on the west coast of India, women are actively involved in marketing, processing, fish drying and curing as well as in net making and mending. While fewer women are involved in such activities in the state of Gujarat, a considerable number of female workers from other states are employed in the fish processing industry of Gujarat.

In India, 50 000 to 60 000 female workers are employed in the fish processing industry alone. Only few female workers employed in fish processing factories are from small-scale fishing communities. The demand for female workers in fish processing factories in the state of Kerala has led to the creation of a category of workers referred to as 'migrant women workers'. These workers are employed on a contract basis through middlemen for seasonal work in fish processing factories in other states. Migrant women workers have been facing problems such as poor working conditions and a lack of medical benefits. The federal government has taken steps to mitigate their problems. In some states, women cooperatives have been established for organizing women for hand braiding of fishing nets, supply of twine and other functions. Commercial banks have extended loans to such societies.

Compared with Southeast Asian countries, where a considerable number of women are involved in aquaculture, involvement of women in aquaculture is very limited in India. The exceptions are family-owned and family-run small-scale fish farms. Women are, however, employed in the collection of wild shrimp and fish seeds for fish farms in some Indian states. They are also involved in shellfish as well as fossil shell collection in estuaries and coastal waters. Dozens of female women workers are employed in each of the 250 shrimp hatcheries in India for packaging of seeds. A few of the ornamental fish farms around Chennai, where ornamental fish is reared for local markets, are run by women.

The main constraints to a wider participation of women in fishery activities are a low level of literacy, social taboos concerning fishing and fisheries related activities, inadequate training and lack of skills as well as a lack of understanding of the socio-economic status of fishing communities at the policymaker level. As part of efforts to alleviate poverty, especially among women, the federal government is promoting self-help groups for fisherwomen as a nucleus for triggering development. Apart from giving training to women in low cost techniques of producing fish products such as dried fish products and fish pickles, financial assistance is extended for setting up fish vending kiosks. Training in low cost techniques for producing various products other than fish is also imparted.

Malaysia

In Malaysia, women are involved in inshore small-scale fisheries activities. In small-scale fish processing establishments as well as industrial processing facilities, women form the major part of the work force. Women are actively involved in small-scale fish processing establishments, which are usually operated at family or household level and where a wide range of traditional products are made. In industrial fish and prawn processing plants including freezing and canning plants, women are employed for dressing, processing, sorting and packaging of fish and work as technologists. Management and supervisory positions are mostly occupied by men.

Many women involved in small-scale fisheries complement their husband's work in marine or freshwater fishing. They are involved in the processing of the part of the daily catch that cannot be sold as well as in other processing and value-adding activities.

In local markets, fish products traditionally processed by fisherwomen are normally regarded as specialties.

There are many success stories of women's involvement in processing of fish-based products all over Malaysia. An example is the production of a fish stick locally known as *keropok lekor,* which is eaten as a snack. More than 90 percent of the cottage industry producers of this home-made snack are women and 98 percent of the workforce involved in producing *keropok lekor* is women. These women entrepreneurs manage to upgrade their families' living status considerably and help to support their families financially.

Women are also involved in marketing of small-scale freshwater and marine catches as well as aquaculture products especially at local weekly markets and at fresh fish stalls. Few programmes and organizations in Malaysia deal specifically with improving the role of women in fishing communities. However, there are several extension programmes and training courses on improved fish processing and freshwater fish culture, which are directed at women. Women also play an active role in fish retailing.

A limited number of women is employed by fishery organizations and involved in fisheries research and extension work. Superstitions that associate women with poor catches, limited financial resources and the absence of technical support services for women are the factors that hinder a greater involvement of women in the fisheries sector. With overemployment already present in the fisheries sector, the government is planning to resettle some fishing communities and train them for employment in other sectors.

Thailand

In Thailand, women are generally involved in fisheries related activities at the subsistence level. In larger fishing villages and towns, women are also actively involved in fisheries related activities at the commercial level, i.e. in fish processing and in freshwater fish culture. Women also work as fish agents and managers of industrial fish and shrimp processing plants and as fishery technologists. Approximately 90 percent of the labour force of fish processing plants is women.

While there are presently no specific programmes for the improvement of the social and economic role of women in fishing communities in Thailand, the general policy of the Department of Fisheries is to improve the social and economic situation of fishermen and their families, to fully utilize the existing labour force including women in the small-scale fisheries sector, to increase the supply of fish, to reduce unemployment in fishing communities and to improve the efficiency and quality of fish processing and fish farming at the village level. In general, a low level of education, conventional taboos regarding women boarding fishing boats and the burden of managing the household, are the hindrances to a greater participation of women in small-scale fisheries.

OPPORTUNITIES FOR DEVELOPING INNOVATIVE SMALL-SCALE FISHERIES PRODUCTS

While there are several opportunities to develop new small-scale fisheries products, there are also constraints like declining profit margins, fluctuations in raw material supply and scarce storage facilities. Apart from fresh fish, the products from small-scale fisheries currently marketed in Malaysia and Thailand are dried fish products, fish jelly products, fermented fish products and fish crackers while in India, fresh and dried product forms still prevail. New opportunities lie in the marketing of ready-to-eat products in the three countries studied as there are an increasing trend in consumption of seafood and an increased awareness of it being a healthy food.

Many new and innovative products can be developed from the fishes landed by small-scale fisheries that would possibly get a better price rather than selling them fresh. Products such as sun-dried, salted and dried, fermented, boiled, marinated, ready-to-eat including breaded and battered products, fish jelly and other products can be produced with simple processing techniques and attractive packaging. These products can be marketed through fish retail outlets, supermarkets and hypermarkets.

As there is already a potential domestic market in the three countries studied, many traditional fish preparations could be reintroduced as specialties to modern consumers. With simple processing methods and attractive packaging, these products could gain popularity in the domestic market, especially with the association of sentimental value and reference to the "good old days" or "grandmother's recipes". Preparation of such traditional products could be done with simple and efficient equipment in order to meet high hygiene and quality standards.

Marketing Opportunities

As a result of the growing global demand for fish and fish products, there are ample opportunities for the small-scale fisheries sectors of the three countries studied to diversify their products and markets. Even though fish landings from small-scale fisheries are scattered, small in volumes and mixed in terms of species caught, through proper coordination and interventions, many of the species landed can be channelled into domestic and export markets in more appropriate value-added forms.

In the case of finfishes, major species of economic importance produced by the small-scale capture fisheries sector in significant volumes are skipjack tuna, yellowfin tuna, sharks, rays, croakers, kingfish, Bombay duck and hilsa (in India only), ribbon fish, seabass and pomfret plus a host of fishes caught by seine nets such as sardines, mackerels and anchovies.

Of the finfish species, tuna species such as skipjack and yellowfin have good potential for value addition. The present range of skipjack products could be further expanded to include improved smoked/dried speciality products such as *katsuobushi* and *arabushi* for the Japanese market, dried and smoked fish flakes and fish extractives. Some of these products are already produced in countries such as Maldives and Sri Lanka for both domestic and export markets. The introduction of improved production and packaging technologies and presentation could further help the small-scale fisheries sector to improve its earnings from skipjack landings.

Yellowfin tuna also holds good potential for product and market diversification. With relatively low capital inputs, the small-scale fisheries sector can embark on producing tuna loins for export and also introduce fresh and frozen steaks and cuts for the domestic market. If proper onboard handling is employed, gilled and gutted yellowfin tuna can also be diverted to export markets as fresh or frozen sashimi tuna. The small-scale fisheries sector can efficiently improve the quality of tuna caught by changing its fishing methods to longlining. This could be achieved with relatively low investments by using smaller fishing boats. Longlining with small boats is increasing in Asia as a result of fuel price hikes and the good quality of tuna landed by these boats, which is in great demand by export processors.

Sharks and rays have good potential for product development as well as for domestic and export marketing. Sharks are the only fishes that can be utilized fully. Fins, skin, meat, teeth, bones and liver are being used for the production of various products. If sharks are properly processed by bleeding the fish onboard, the meat could be marketed domestically in chilled form and exported in frozen form. Though there is a domestic demand for shark fins in Thailand and Malaysia, which caters to Chinese restaurants, most of the local production is exported. Ray wings in frozen and dried form have a potential market in China, Hong Kong SAR and China.

Pelagic fishes like sardines, mackerels and anchovies are caught by traditional fishermen using local gears. These fishes could be processed into different products, which have a ready market both domestically and for export. Dressed and marinated mackerels and sardines have a potential market in the Middle East while boiled mackerels have a ready market in Singapore, Brunei and the Philippines. If anchovies could be better handled onboard, they could be used for the production of boiled dried anchovies, which have a ready market in Japan and South Korea.

Thailand and Malaysia

Tilapia is produced by both capture and culture fisheries in inland areas. Tilapia is one of the most sought after fishes both in the international and domestic market, especially in Malaysia and Thailand. The demand for tilapia products picked up in Thailand and Malaysia after the economic recession of 1997 and 1998. Currently, tilapia is being marketed domestically in live and fresh/chilled form in Thailand and Malaysia and a small portion of the production is being exported to nearby markets like Singapore in live form. There is an opportunity to market fresh/

chilled and gutted whole tilapia with improved packaging like tray packs in these markets through super- and hypermarkets as the price difference between wet markets and super- and hypermarkets is more than 20 percent. Chilled and frozen tilapia fillets also hold good market potential in both the domestic and export market. With small capital investments, chilled whole tilapia and fillets could be produced by the small-scale fisheries sector for export. Marinated tilapia fillets also have a ready domestic market.

Other commercially important finfish varieties landed by the small-scale fisheries sector in the three countries studied are kingfish or Spanish mackerel, seabass, grouper and pomfret, which have a ready domestic and export market. These fishes are currently marketed domestically in fresh/chilled form. There is a good export market for products such as slices, steaks and fillets of these fishes in fresh/chilled and frozen form.

Annually, about 44 000 tonnes of hilsa (*Hilsa illisha*) are landed in India. The entire production is consumed locally in fresh form. There is a potential for marketing dressed or sliced hilsa with improved packaging domestically and there is a ready market for frozen hilsa in markets like the Middle East and Malaysia. As regards Bombay duck (*Harpodon nehrerus*), a unique fish occurring in Indian waters, about 165 000 tonnes are landed annually. Almost the entire catch is consumed domestically in dried form. There is an opportunity to market dried and laminated Bombay duck in Europe, especially in the UK. The dried and laminated Bombay duck could also fetch a much better price than the normal dried product in the domestic market as the product has a much longer shelf-life and a better appearance.

Significantly large quantities of carps are produced in all three countries studied. Currently, all carps are marketed domestically in fresh or chilled form except for some quantities exported from India to nearby export markets. There is a promising market for fresh/chilled and frozen carps and value-added carp products such as carp cooked in traditional curry sauces.

All three countries studied produce ornamental fishes through marine and freshwater capture fisheries and aquaculture. India and Malaysia presently do not export ornamental marine fishes.

Their exports mainly consist of freshwater fishes both from aquaculture and capture fisheries. The fishing methods used are entirely artisanal. The ornamental fish culture sector consists exclusively of small-scale enterprises. Ornamental fishes cater to both the domestic and export markets. Opportunities for this sector lie in the introduction of new species from capture fisheries and modified species from the aquaculture sector. Most of the ornamental fishes produced by capture fisheries and aquaculture have export potential.

The small-scale fisheries sector produces large volumes of shellfish such as shrimp, crab and freshwater prawn both through capture fisheries and aquaculture. About 85 to 90 percent of the shrimp produced by shrimp farming is being exported by India, Malaysia and Thailand. The present range of shrimp products could be further expanded to include value-added products such as head-on shrimp, peeled tail-on shrimp, seafood mix composed of squid, cuttlefish, mussels/clam and shrimp and breaded and battered products both for domestic and export markets. Value-added shrimp products have a very good potential in international markets. While freshwater prawn from Thailand and Malaysia is marketed in fresh/chilled whole form in the domestic market leaving only a limited quantity for export, almost the entire production from India is exported. There is a potential for processing of freshwater prawn in head-on and headless deep-cut forms for the export market, which would have price advantages.

Crab is marketed domestically in live and fresh/chilled form. There is a large potential for value-added crab products both for the domestic and export markets. The products identified for value-added production are cut crab (half-cut and quarter-cut crab), stuffed crab, crab balls, stuffed claws, picked meat and crab mince meat products. These products have also potential in the domestic market, particularly when marketed in retail packs.

A large quantity of molluscs of economic importance is produced by the small-scale capture fishery sector such as squid, cuttlefish and octopus, which are harvested mainly by artisanal fishers using hook and line. Bivalves such as cockles, mussels and clams are also harvested by small-scale fishers by dredging. The present product forms of squid, cuttlefish and octopus marketed by the small-scale fisheries sectors in the domestic markets of Thailand and Malaysia are fresh/chilled whole products and dried products while there is only a small demand for cephalopod products in the Indian domestic market. The potential products that could be developed with improved packaging are fillets (pine-cut, shell-cut and double-skinned), squid rings and cuttlefish strips, which have a potential market both domestically and for export.

Bivalve products such as cockles, mussels and clams have a good potential for value addition. Currently, bivalve products are marketed in the form of boiled and chucked meat. This could be further expanded to include improved smoked/dried specialty products and live products for export to Japan and Singapore. Introduction of improved production methods with innovative packaging and presentation could also improve earnings.

In India, Thailand and Malaysia, bycatch constitutes a sizable portion of the total catch, i.e. 20 to 40 percent. Much of the bycatch consists of low value pelagic fishes. Most of these are currently utilized for the production of fishmeal. Fishmeal is used as poultry feed and the production of aquaculture feed. Almost all so-called trash fish landed at present is used for this purpose.

A smaller portion of bycatch is used for the preparation of dried, salted and fermented fish products. In Malaysia and Thailand, bycatch is also used to prepare fish jelly products and fish and shrimp crackers.

In Thailand and Malaysia, there are 1 215 and 136 processing plants, respectively, which produce these types of products. In India, selected bycatch species such as croakers, threadfin breams and goat fishes, which have white meat, are used by nine commercial surimi processing plants.

With regard to traditional products, dried fish processing is one of the most widespread small-scale industries in the countries studied. Depending on seasonal availability, a large amount of species are used for the production of dried fish. Among these, the most commonly used species are sardines, mackerels, ribbon fish, lizard fish, sharks, threadfin bream and snappers. Most of these fishes are either sun or kiln dried with a moisture content of up to 40 percent. The polyethylene bag packed products are marketed through retail stores, open markets and night markets in urban and suburban areas. Improved product processing and presentation are the opportunities available for this sector.

Fish and shrimp crackers are one of the most popular snack foods in Malaysia and Thailand. Fish or shrimp crackers are not produced in India yet. Most of the fish and shrimp crackers are produced by small-scale operators from low value fish species or small shrimp. Species like Japanese threadfin bream, bigeye tuna and goat fish are commonly used. The main ingredients are fish and shrimp mince, tapioca flour, salt and spices. There are good opportunities to produce fish and shrimp crackers with improved technologies and innovative packaging and presentation as there is a great demand for this product in domestic and export markets.

Minced fish products such as surimi, fish balls and fish cakes are favourite snack foods in Thailand and Malaysia. Fish balls are one of the ingredients in soup preparations and other dishes in these two countries.

The ingredients of fish balls and fish cakes are fish mince, tapioca, corn starch, salt, baking powder and various seasonings. In Malaysia, there are over 28 fish ball, fish cake and surimi processing facilities. Thailand has 95 facilities. While the majority of these facilities are operated by the small-scale sector, there are also a few commercial establishments with sophisticated machinery. The products are marketed chilled and sold in wet markets, super- and hypermarkets. There are opportunities for export of these products to regional markets.

In Malaysia and Thailand, fish sauce products are traditionally produced in small-scale operations. Fish sauce is an indispensable part of many Thai and Malaysian dishes. The basic process of fish sauce production involves fermentation of pelagic fish species such as anchovies, sardines and mackerels in salt. Fish sauce made of anchovies generally commands higher prices than fish sauces made from other fishes. Freshwater finfish varieties such as eel, catfish, snakehead and barbs are currently being used for the preparation of a number of traditional fish products. With low capital inputs, the small-scale fisheries sector can embark on the production of traditional fish products from freshwater fishes for export to regional and ethnic markets.

Improved Utilization of Catch, Bycatch and Waste

Very little bycatch is discarded by the small-scale fisheries sector in Asia. Most of the marine and freshwater species caught by small-scale fisheries are traditionally consumed in fresh form in all three countries studied including bycatch species. These species have been used for centuries for the preparation of a wide range of traditional fish products such as salted and dried fish, fish sauce and fish paste. The introduction of modern technologies has significantly increased the quality and widened the range of products. This has resulted not only in the rapid commercialization of the traditional processing sector but also in the introduction of new products based on bycatch species such as surimi, a wide range of fish jelly products, fish cake and fish floss.

With a view to promote a sustainable use of fishery resources, it is important to preserve the quality of fish and bycatch on board small-scale fishing vessels to facilitate an improved utilization of catches for human consumption. It might not always be possible to achieve improved onboard preservation and handling of catches due to various limitations such as a lack of storage space for ice and bycatch. In Thailand, attempts have been made to convince fishers to use ice for the preservation of bycatch and trash fish with the purpose of assuring them of a better price for these products. Infrastructure at landing sites also plays an important role in improved bycatch utilization. Availability of clean water, ice for re-icing of catch if necessary, storage space, mechanical facilities for unloading bycatch without damaging it, transportation facilities as well as wholesale and retail facilities and services are important in this respect. Much progress has been made in the region with the utilization and marketing of fish waste, especially shrimp waste. Production of chitin and chitosan and extraction of astaxanthin and other pigments and feed additives have become an integral part of the shrimp industry in most countries in the region. Government-industry cooperation in promoting such activities by way of technology transfer, market promotion and streamlining the collection of waste from processing centres would be helpful in promoting waste utilization further.

Product and market diversification is another important aspect to be addressed for improving the utilization of bycatch and mixed catch. The fish products, which are presently being marketed could be further improved by extending their shelf-life, safety and quality. Modern methods of packaging, transportation and storage could be used to market some species of bycatch in dressed forms and in consumer packs or in a ready-to-eat/prepare form. Over the last decade, the number of small-scale bycatch processing enterprises has declined due to competition from large-scale operators with capital and advanced technological know-how. Transfer of such technologies to the small-scale fish processing sector, related training and improving entrepreneurship could be considered as important steps to improve bycatch utilization and marketability in the small-scale fisheries sector.

Assurance of product quality and safety is a vital step in the marketing of fish products. Considering the diffused nature of raw material supply in the bycatch processing sector and the traditional nature of the production technologies involved, it is important to have HACCP based production systems in place to ensure that fish products gain consumer confidence. In this respect, support and assistance to the small-scale fisheries sector should be provided by regulatory authorities and governments.

Governments in the three countries studied have a general policy to promote small-scale entrepreneurship by providing credit and financing facilities for the agriculture and fisheries sector through agricultural banks. In many cases, banks offer favourable terms and conditions through government assisted schemes. Many small-scale operators, however, prefer to obtain funding from sources outside the formal banking system, even though higher interest rates are being charged as they find the government assisted financing schemes operated through the banking system too complicated because of the paperwork involved and the collateral required. It would be prudent to simplify such assistance packages and, wherever possible, to channel them through specialized fisheries and cooperative banks, which are able to offer user-friendly schemes.

Even though the demand for bycatch is expected to continue in the years to come, many doubt the long-term availability of fishery resources for this purpose as all governments in the region have initiated action to minimize the catch of non-target species.

Considering conflicting interests among stakeholders, a successful implementation of these measures becomes difficult.

Traditional approaches to improve bycatch utilization need to be reassessed in the light of new developments in the global market. In the past, most traditional products, which are the key bycatch based products of today, were processed by small-scale cottage industries with very little mechanization. Present-day product processing and marketing has to address special criteria not only with respect to quality and safety of products but also with respect to packaging, labelling including ecolabelling, presentation and resource sustainability. It is also important to have a good knowledge of market requirements and consumer expectations including ready access to market and price information. In such a scenario, human and material requirements and costs of fish processing and marketing are increasing in terms of trained labour, market and product information, upgraded processing facilities, quality control and monitoring.

STRATEGIES FOR OVERCOMING CONSTRAINTS TO THE DEVELOPMENT OF SMALL-SCALE FISHERIES

Constraints to the Development of Small-scale Fisheries

The following constraints to the development of small-scale fisheries and the full integration of the sector into fish trade prevail in India, Thailand and Malaysia.

India

Boats typically used in the small-scale fisheries sector, especially non-powered small boats and boats used for single day fishing, are not equipped with facilities for preserving fish. Inadequate facilities for preservation of fish, unhygienic conditions at landing centres, lack of supply of quality ice, inadequate transport facilities, lack of cold chains, lack of modern fish markets and a large number of intermediaries involved in marketing are the major constraints.

Another constraint is the fact that carp dominates the bulk of production from freshwater aquaculture while shrimp is the only item produced by brackishwater aquaculture. As far as product development is concerned, no significant efforts are being made to develop value-added products for domestic or export markets. Government taxes and duties pose a further constraint. Processed or frozen seafood products attract excise duty of 8 percent plus value-added tax. This system imposes higher charges on the fish processing sector in India, which is not seen in other countries in the region.

There is a lack of participation of women in fish marketing and processing caused by low levels of literacy, social taboos, inadequate training in skills development, a low socio-economic status of traditional fishing communities and underemployment of fisherfolk due to fishery resource limitations.

While there is a huge potential for the development of the domestic fish marketing in India, efforts are not made to tap this potential. Emphasis is given to export marketing instead. The nutritional value and health benefits of fish are not widely known. Consumers are sceptical of frozen fish products and do not know how to use them.

Malaysia

As the volume of fish handled by the small-scale fisheries sector is relatively small compared with the commercial sector, the maintenance of regular supplies of raw materials to plan fish processing activities is a major constraint. Small-scale fisheries operators also need help to acquire appropriate and low cost fish processing technologies suited to their production levels and standards and to their desired products.

Another constraint is posed by inadequate storage facilities and a lack of knowledge of new fish products. Small-scale fishers have limited storage facilities onboard their vessels as well as onshore for preserving their catches properly. They further lack knowledge on the development of new products and on quality and safety aspects of fish products.

Lack of finance is another crucial constraint faced by the small-scale fisheries sector in Malaysia. Although there is an exclusive financial institution with the objective of financially assisting small-scale entrepreneurs including fishers, the institution is inefficient in reaching small-scale fishers, who are often not aware of the services and facilities offered by the institution.

Thailand

The coastal waters of Thailand are overfished and conflicts between different groups of resource users for scarce fishery resources are common. Fishery resources depletion and a rapid urbanization and industrialization in coastal areas causes changes in the patterns of land use, pollution of coastal waters and the degradation of the coastal and aquatic environment. These changes particularly affect the small-scale fisheries sector negatively. The same applies to inland fisheries. The problems are amplified by the increasing costs of production particularly due to increases in fuel costs and a low price realization with regard to some species caught. Shortage of labour in commercial fisheries is another problem.

Regarding aquaculture, the problems encountered by shrimp farmers include many types of diseases, which affect cultured fish and shrimp species, environmental degradation caused by inefficient farm management, mangrove area encroachment, insufficient natural broodstock and increasing costs of farming operations. Other constraints encountered by small-scale fishers, fish and shrimp farmers and fish processors are a lack of capital, a lack of fish and shrimp farming and fish processing experience and a lack of knowledge of suitable fish and shrimp farming and fish processing technologies and methods. Small-scale fishers and fish and shrimp farmers have limited bargaining power and face stiff competition from fish producers in neighbouring countries.

STEPS FOR THE DEVELOPMENT OF SMALL-SCALE FISHERIES

To overcome the abovementioned constraints and problems, the following steps can be taken.

India

To improve infrastructure facilities and the effectiveness of marketing channels, efforts should focus on improving facilities for preserving fish onboard vessels and hygienic conditions at fish landing centres, increasing the supply of quality ice, establishing cold chains and marketing systems, streamlining marketing channels by strengthening the role of wholesalers and reducing the number of intermediaries. A proper management system for fish landing centres should be introduced for their day-to-day maintenance and generation of income to meet maintenance expenditure. The management of such facilities could be handed over to fishermen cooperative societies or *panchayats* in villages. All major landing centres in inland areas should be provided with the necessary infrastructure for auctioning and packaging of fish.

Fishermen cooperatives should be strengthened to undertake marketing of fish so that the share of fishers of the price paid by the final consumer is maximized. Private associations of fishermen should be encouraged to undertake marketing of their products on a larger scale. Modern fish markets should be established in major urban centres. This will not only provide ideal conditions for selling fish but also boost the image of fish as a quality food among consumers. Development of a legal framework for the establishment and management of fish markets is necessary to supply safe and quality fish products to consumers.

Culture of suitable new species of fish such as Nile tilapia should be encouraged to meet the demand for fish among consumers. The development of value-added products from low value fish species should be urgently promoted. If necessary, foreign expertise should be sought. In order to promote the marketing of frozen fish products, the excise duty on these products should be waived. Value-added tax (VAT) should also be reduced.

Women self-help groups should be promoted to encourage women to engage in income generating fisheries activities such as the preparation of value-added fish products and their marketing. Proper training in the techniques of production and marketing should be provided.

Fish as a healthy food needs to be popularized among consumers. A special campaign to promote eating of fish is necessary similar to the campaign currently undertaken for dairy products and poultry. It is necessary to formulate a nation-wide fish marketing strategy with the specific objectives of helping fishers to market their products at a remunerative price and to supply safe and quality fish and fish products to consumers. The example of the cooperative structure of the small-scale dairy industry in India should be followed.

Last but not least, the improvement of fishery statistics, especially statistics regarding inland fisheries landings and marketing of fish from various sources, is necessary for better planning and management of the use of fishery resources and fish marketing.

Malaysia

The inconsistency of supply of raw material for small-scale fish processing should be addressed by identifying proper storage methods to ensure more regular and higher incomes for small-scale fish processors. In addition, seasonal products should be introduced, which would fetch higher prices.

Storage facilities to suit the scale of fish handling, preservation and processing operations should be promoted such as the use of insulated fish boxes for preserving catch onboard vessels and the use of chest freezers to store and preserve fish onshore before it is processed. Small-scale fishers, fish farmers and fish processors should also be provided with knowledge of quality aspects of fish and fish products and of the development of new products and a more hygienic production of traditional products.

Financial assistance and credit facilities should be extended through local area fishermen associations of the LKIM (Fisheries Development Authority of Malaysia), the Department of Fisheries or through the Bank Pertanian Malaysia – Agriculture Bank of Malaysia (BMP) as appropriate.

Thailand

In order to overcome the conflicts among the different users of Thailand's coastal fishery resources, clearly demarcated fishing areas or zones should be allocated to different resource users and the government should strictly implement its regulations in this respect.

In addition to a general shortage of raw material, other problems in the fish processing sector are the limited volume of products available for local consumption and the high marketing costs. Government agencies should study this problem and formulate an action plan to address it. Government agencies should further assist in identifying the problems in the aquaculture sector mentioned above and suggest remedial measures. Government agencies and training institutes should also impart training on various aspects of fish and shrimp farming including modern farming techniques. Financial institutions should work hand in hand with the Department of Fisheries of Thailand and other related agencies to extend credit and microfinance and other financial assistance to the small-scale fisheries sector.

The export of traditional fish products from Thailand is increasing and has a very good further potential. In 2004, Thailand exported over 55 000 tonnes of traditional products valued at US$54.7 million. Ways and means of further increasing these exports of traditional fish products should be explored.

Small-scale Fisheries and Fish Trade in Western Africa

SMALL-SCALE FISHERIES AND FISH TRADE – THE CASE OF GHANA

Ghana has a coastline of about 550 km. The Exclusive Economic Zone (EEZ) waters extend up to 200 nautical miles from the shore. The fishery industry of Ghana comprises the marine sector and the inland sector. The marine fisheries sector is the main source of fish producing 85 percent of the total catch. The inland sector accounts for the remaining 15 percent of Ghana's fisheries production exploiting a large number of rivers, irrigation dams, ponds and lakes including Lake Volta, one of the largest man-made lakes in the world, with a total surface area of 8 700 km^2.

Marine capture fisheries production and resources

In 2001, about 123 000 artisanal marine fishermen operated 10 000 dugout canoes from 304 landing centres located in 189 fishing villages. In addition, there were 296 beach landing sites located outside of villages. The semi-industrial inshore fishing fleet consists of locally built wooden vessels of 8 to 37 m in length. These vessels are primarily used for sardinella fishing and trawling. The number of vessels has been decreasing over the past 15 years due to a decline of target species and increasing costs of operation and maintenance.

In the artisanal fishery sector, several types of fishing gears are used including a wide variety of gillnets, entangling nets, purse seines, beach seines, handlines and castnets. There are fisheries for small pelagic species of the families Clupeidae (sardinellas), Scombridae (mackerels) and Engraulidae (anchovies) and for large pelagic species of the family Thunnidae (tunas). Fishing for sardinella (*Sardinella aurita* and *Sardinella maderensis*) is one of the most important economic activities in the Ghanaian fishing industry.

Large variations in landings of sardinella are experienced from year to year due to changes in both anthropogenic and natural factors.

The former are exemplified by changes in fishing effort and fishing strategies and the latter by changes in the duration and intensity of the seasonal coastal upwelling. The abundance of chub mackerel (*Scomber japonicus*) also varies from year to year. The potential yield of the four most important small pelagic species, i.e. round sardinella, flat sardinella, chub mackerel and anchovy is about 200 000 tonnes per annum.

The main commercial tuna species, which occur in Ghanaian waters, are yellowfin tuna (*Thunnus albacares*), skipjack (*Katsuwonus pelamis*) and bigeye tuna (*Thunnus obesus*). Recent assessments undertaken by the International Commission for the Conservation of Atlantic Tunas (ICCAT) indicate that yellowfin and bigeye tuna resources in the Atlantic are being optimally exploited while skipjack tuna is underexploited. The total landings of the three major tuna species by Ghanaian fishing vessels are about 60 000 to 80 000 tonnes per annum.

There are also important fisheries for demersal species of the families Sparidae, Lutjanidae, Mulidae, Pomadasyidae, Serranidae, Polynemidae and Penaeidae. The potential yield of demersal fishes on Ghana's continental shelf is estimated to be about 55 000 tonnes per annum. During the last decade, annual landings averaged about 50 000 tonnes.

Tables 1 to 4 show the catches of the artisanal and semi-industrial fishing fleets, i.e. inshore purse seiners and trawlers. While the catch of the artisanal marine fishing fleet in 2002 was 200 769.19 tonnes, the semi-industrial fleets only caught 7 784.55 tonnes.

Table 1: Artisanal marine fisheries production by canoes in Ghana, 1996-2002 (in tonnes)

Species/y	1996	1997	1998	1999	2000	2001	2002
Roun d	115 069.8	46 383.93	54 595.63	39 123.91	98 864.69	64 103.56	59 399.56
Flat sardin	13 501.30	14 069.74	14 770.28	12 055.71	14 934.93	15 853.37	13 693.40
Chub mack	6 855.80	7 086.09	1 971.95	5 240.61	14 835.38	9 574.30	5 428.90
Anchovy	98 340.50	82	44	32	83	68	57
Friga te	7 952.60	5 655.37	7 326.36	4 209.42	3 627.66	7 323.50	3 635.90
Seabream	9 571.70	8	13	19	1	7	4 741.05
Burrito	9 706.00	17	9	11	8	12	7 573.20
Others	37 251.30	33	42	5	50	51	48
Total artisanal productio	298 249.0	215 125 .44	189 458. 60	176 236. 88	275 964 .69	236 355 .26	200 769.19

Source: Directorate of Fisheries, Ghana, April 2004

Table 2: Marine fisheries production by semi-industrial inshore purse seine vessels in Ghana, 1996-2002

(in tonnes)

Species/year	1996	1997	1998	1999	2000	2001	2002
Round sardine	3 330.50	3 009.65	1 369.41	1 255.51	3 177.99	3 208.81	3 449.14
Flat sardine	117.81	113.32	697.84	49.09	34.97	529.79	80.91
Chub mackerel	1 115.00	1 054.88	773.32	1 369.64	3 630.21.	971.81	891.06
Scad mackerel	45.00	4.04	334.64	42.58	5.14	119.46	149.93
Others	1 243.00	738.44	399.15	785.54	368.21	381.89	403.26
Total	5 851.31	4 920.33	3 574.36	3 502.36	7 216.52	5 211.76	4 974.30

Source: Directorate of Fisheries, Ghana, April 2004

Table 3: Marine fisheries production by semi-industrial inshore trawlers in Ghana, 1996-2002 (in tonnes)

Species/year	1996	1997	1998	1999	2000	2001	2002
Seabream	30.20	33.39	37.62	19.90	27.47	266.29	70.80
Cassava fish	293.90	330.30	392.69	303.46	255.03	425.76	524.48
Burrito	1 196.80	900.52	564.17	518.72	450.35	632.08	679.38
Trigger fish	0.20	0.22	0.00	0.18	-	-	-
Red mullet	2.80	2.14	40.15	0.27	0.12	2.35	3.02
Flying Gurnard	67.70	67.71	0.40	0.06	-	-	-
Cuttlefish	13.20	14.05	35.29	62.47	63.15	56.89	40.21
Others	896.70	1 024.62	1 532.71	741.95	655.42	1 010.41	1 492.36
Total	2 501.50	2 372.95	2 603.03	1 647.01	1 451.54	2393.78	2 810.25

Source: Directorate of Fisheries, Ghana, April 2004

Inland capture fisheries production and resources

Fishing in Lake Volta accounts for about 90 percent of the total inland fish production in Ghana. Formed about 40 years ago, Lake Volta with a shoreline of about 5 200 km is the largest man-made lake in Africa. After the creation of the lake, fishers from various parts of Ghana moved to the lake area. At present, 80 000 fishers and 20 000 fish processors and traders are engaged in the Lake Volta fishery, which is of an artisanal nature. The fishing gears used are castnets, gillnets, hooks and lines and traps operated by 17 500 planked canoes. During the initial years, over 100 fish species were found to inhabit the lake. Perch species (*Cichlids)* accounted for about 50 percent of the catch. Perch species are still the most common fish species landed.

The top ten freshwater species in terms of landings are *Tilapias* (38.1 percent), *Chrysichthys* spp. (34.4 percent), *Synodontis* spp. (11.4 percent), *Labeo* spp. (3.4 percent), *Mormyrids* (2 percent) and *Heterotis* spp. (1.5 percent). Other species of commercial importance are *Clarias* spp. and *Bagrus* spp. Significant seasonal and annual variations are recorded in the fish production from inland sources.

There are more than 50 lagoons of various sizes located in the coastal areas of Ghana. These lagoons provide an important source of protein and other resources for the communities that live around them. The lagoons also contribute significantly to the biodiversity and status of fish stocks in coastal waters as many fish species spend a part of their life cycle in these lagoons. Currently, these lagoons and their ecology and environment are degraded by human activities, i.e. physical alterations and destruction of habitat as a result of changing land use patterns and pollution. The mangrove forests that once fringed many of the lagoons have been lost. Consequently, fisheries in the lagoons are threatened by environmental degradation and destructive fishing practices.

Fisheries in the lagoons are solely artisanal and play an important role in the economy of some coastal communities. The fishing gears used in lagoon fisheries include castnets, dragnets and various types of traps. Tilapias are the most abundant species in all lagoons while a number of freshwater and marine fish species are also caught. These include mud fish, bonga and sole. Some marine species like *Lutjanus fulgens* (snapper), *Caranx hippos* (jack mackerel) and *Epinephelus aeneus* (groupers) only make short incursions into the lagoons.

Aquaculture

Even though fish farming is new to Ghanaians, its practice is becoming more widespread in the country, especially in the Ashanti, Brong Ahafo, central, eastern, Volta and western regions. Fish farming has become an option for increasing the fish production in Ghana since the marine and inland capture fisheries production has reached its maximum sustainable level.

Table 4: Inland fish farming production in Ghana, 1998-2000 (quantity in tonnes, value in US$1 000)

Species	1998		1999		2000	
	Quantit	Value	Quantit	Value	Quantit	Value
Heterotis niloticus	n/a	n/a	n/a	n/a	19	30.1
Catfish	105	210	110	220	76	120.4
Nile tilapia	315	519.8	320	528	347	687.1

Source: FAO FISHSTAT, 2000

There are about 1 000 fish farmers in Ghana working on over 2 000 ponds with a total surface area of 350 ha. Both extensive and semi-intensive aquaculture methods are practised. Extensive aquaculture is associated with the use of dugout canoes and small reservoirs, which are fished and restocked. Fish is cultured semi-intensively in earthen ponds either as monoculture or polyculture of tilapia (especially *Oreochromis niloticus*) and catfish. Cage and pen culture are practised in lakes, lagoons and rivers.

Fish handling and preservation

Smoking, salting, drying, icing, cold storage and canning are used for preserving fish in Ghana. The traditional methods of smoking, salting and drying are used to preserve most of the fish from the artisanal sector and the semi-industrial inshore fleet. In Ghana, as in other parts of West Africa, about 80 percent of the fish is consumed smoked and the remaining 20 percent is consumed either fresh, salted, sun dried or fried. The type of fish determines to a large extent the method used to preserve it. This is again related to the preference of consumers.

Smoking is used for most fish species and is carried out in almost all fishing villages. Wood is therefore an important requirement in these areas. Traditional smoking equipment such as the round mud oven has been replaced in most fishing villages by rectangular ovens with smoking trays. These so-called Chorkor smoking ovens have a higher smoking capacity because they use stacks of smoking trays. They also consume less fuel than the traditional ovens for the same quantity of fish. Sun-drying is used to dry small-size species like anchovies. Generally, fish is dried on mats, cemented floors, grass, old fishing nets and sometimes directly on the ground. Ice is used to extend the shelf-life of fish landed by the semi-industrial inshore vessels, which mostly carry ice on board. Some canoe fishermen called *Lagas,* who use hook and line for high value demersal species, also carry ice to sea in insulated containers stored in special compartments in their canoes. Keeping certain fish species in ice after capture can be logistically difficult and uneconomical for artisanal fishermen. Pelagic species such as sardinella and mackerels do not command high prices and therefore will not attract higher prices when kept in ice.

Fish Marketing

Pricing

Prices of small pelagics fluctuate widely throughout the year and across the country.

Table 5: Prices of smoked and dried fish by species, Accra, Ghana, 2001 (in GHC)

Product forms	Average wholesale prices in GHC[4]/kg
Smoked sardinella	400
Smoked anchovy	200
Smoked tilapia	1000
Smoked African perch	1200
Dried anchovy	200
Salted and dried tilapia	750
Fermented tilapia	800

Source: Salago Market, Accra, 2001

During the peak fishing season when supply increases, prices are lower than during the lean season when supply is lower than demand. Fish prices also increase as the distances from the landing sites and catch areas increase.

The prices shown in Table 5 were recorded at the Salaga market in Accra from semi-wholesalers, who sell their products to retailers. When smoked fish products are destined for foreign countries, prices are set according to the size of the traditional baskets made with fresh palm leaves. The quantity varies from 30 to 60 kg. Smoked fish products are also exported in plastic sacks, which weigh up to 100 kg. Table 6 shows the average prices of smoked, salted and dried fish products in Ghana.

A modern domestic market for fish products in Ghana is not yet developed. Ghana supplies its eastern neighbours, i.e. Togo and Benin with considerable volumes of freshwater fish, especially smoked tilapia, smoked catfish and salted and dried tilapia from Lake Volta.

Trade in processed small pelagic species such as smoked sardinella, smoked anchovy and dried anchovy is gaining prominence. Even outside the main fishing seasons, 400 to 600 baskets of cured small pelagics weighing about 60 kg each cross the Ghana-Togo border each week originating from supply centres such as Mamprobi (Tuesday Market, Tema, Keta, and Denu). During the peak season for small pelagics from June to October, trade volumes double or even triple making the Ghanaian trade in cured small pelagics one of the most important cured fish trade activities in the region. The volume of Ghana's cured small pelagic trade is estimated between 1 500 and 3 000 tonnes per year with smoked sardinella constituting the bulk of the trade. The main supply source is the Tuesday Market in Mamprobi, Accra.

As the name implies, markets are held every Tuesday on a weekly basis, with buyers and sellers converging from within and outside Ghana to transact business. The buyers come from Togo, mainly from the Hutokpamé market, and from Benin, mainly from the Dantopka market, as well as from other parts of Ghana. The sellers bring their stocks of cured fish from the vicinity of the Accra metropolitan area including supply centres such as Akplabanya, Botianor, Chorkor, Nyanyano and Teshie.

Apparently, most of the transport operators engaged in this activity are based in Dzodze and therefore prefer to use the following trade route: Mamprobi/Accra – Sogakofe – Dzodze/ Ghana – Noepé – Sanguera – Lomé/Togo. Another convincing explanation for the choice of this trade route is that the Dzodze border post is less busy than other border posts and customs formalities proceed much faster. Some transport operators use the more direct route through Denu and Aflao but they constitute a minority since the Aflao border post is very busy. Few traders supply the interior markets in Togo notably Atakpamé. Such traders use the following trade route: Mamprobi/Accra – Hohoe – Kedjebi/ Ghana – Badou – Atakpamé/Togo.

Women's contribution to fish trade

Throughout the world, the fishing industry has long been considered a male occupation and many international agencies and business development programmes are solely focussing on fishermen. But in Ghana, women play an important role in small-scale fisheries. The wholesalers, locally called fish mammies, handle distribution and sales of 90 percent of the artisanal fish production as well as part of the catch of the commercial fishing companies. Most of the fish mammies operate as selling agents paying fishermen on the basis of the revenues they receive from their sales. Fish mammies also provide informal credit to fishermen and are involved in processing, i.e. smoking, drying and salting of fish for the domestic market as well as for regional markets.

SMALL-SCALE FISHERIES AND FISH TRADE – THE CASE OF SENEGAL

Since 1994, when the CFA Franc (CFAF) was devaluated, fish products became the main export product and foreign exchange earner in Senegal replacing groundnuts, phosphates and tourism. At present, the fishing industry accounts for 30 percent of Senegal's export earnings and is the most important foreign currency earner. Within the fishery sector, the small-scale or artisanal fisheries sector plays an important role.

The artisanal fisheries sector, including those involved in fish processing, marketing, net making, boatbuilding and similar activities, employs 600 000 Senegalese or 17 percent of the working population of Senegal. The artisanal fisheries production contributes nearly 70 percent to the domestic fish consumption.

Small-scale marine and inland capture fisheries production and resources

Small-scale fishing is done with small motorized and non-motorized wooden boats and dugout canoes. Ninety-one percent of the fishing canoes are motorized. Seventy-three percent of the motorized fishing canoes operate from Thiès and 9 percent from Dakar. Table 3.7 shows the number of canoes during the period 2000 to 2004.

Table 7: Canoes operating in Senegalese territorial waters

Year	Number of canoes
2000	6 329
2001	7 910
2002	8 351
2003	8 557
2004	6 050

Source: Annual Report, Directorate of Marine Fisheries (DPM), 2000

In 1997, 186 fish landing sites were recorded. The registered number of marine and inland fishermen was 53 000. In 2000, the total production of the small-scale fisheries sector was 338 209 tonnes. The main species exploited by the small-scale fisheries sector are two types of sardinella, i.e. *Sardinella aurita* and *Sardinella maderenis* and horse mackerels. The total small pelagic resources of the Senegalese EEZ are fully exploited and overexploited even though there seem to be some regional differences. In the Casamance region in the south, there may be potential for a slight increase of production of sardinella as well as potential for catching of bigeye grunt (*Brachydeuterus auritus*), which is presently hardly exploited, whereas fishery resources off the Petite Côte region are overexploited.

Artisanal inland capture fisheries is mainly carried out in rivers using canoes, gillnets, castnets, traps and hook and lines.

The artisanal marine and inland fisheries production during the period 1997-2002 is shown in Table 3.8 and 3.9.

As can be seen from Table 8, the artisanal marine capture fisheries production in Senegal shows an overall decline,. which is probably due to overfishing of the available resources.

Table 8: Small-scale marine fisheries production of Senegal, 1997-2002 (in tonnes)

Year	1997	1998	1999	2000	2001	2002
Quantity	345 600	317 100	302 300	328 800	320 400	292 000

Source: Ministry of Maritime Economic Affairs (MEM) & Directorate of Statistics (DPS), 2000

Table 9: Small-scale inland fisheries production of Senegal, 1997-2002 (in tonnes)

Year	1997	1998	1999	2000	2001	2002
Quantity	23 800	23 600	27 400	25 300	27 700	35 200

Source: Ministry of Maritime Economic Affairs (MEM) & Directorate of Statistics (DPS), Ministry of Economic Affairs and Finance (MEF), 2000

Inland capture fisheries landings in Senegal show an overall increase even though at a low level. The situation is mainly due to the economic recession in Senegal as well as successive years of draught, the drying up of several rivers and the modification of other rivers for irrigation schemes.

Aquaculture

Fish farming in the form of Nile tilapia freshwater culture was introduced in Senegal in 1980. In 2000, the production was 204.8 tonnes valued at US$50 000. The production is destined for the domestic market in fresh form. The price is set according to the size of the fish and the period of production.

Fish handling and preservation

As in the case of Ghana, smoking, salting, drying, icing, cold storage and canning are used for preserving fish in Senegal. The traditional methods of smoking, salting and drying are used to preserve most of the fish from the artisanal sector and the semi-industrial inshore fleet. In Senegal as in other parts of West Africa, about 80 percent of fish is consumed smoked and the remaining 20 percent is consumed either fresh, salted, sun dried or fried. The type of fish determines to a large extent the method used to preserve it and this again is related to the preference of consumers.

Smoking is used for most fish species and is carried out in almost all fishing villages. Wood is therefore an important requirement in these areas. Traditional smoking equipment such as the round mud oven has been replaced in most fishing villages by rectangular ovens with smoking trays. The so-called Chorkor smoking oven has a higher smoking capacity because it uses stacks of smoking trays. It also consumes less fuel than the traditional ovens for the same quantity of fish. Most of the fishing villages have adopted the use of the Chorkor oven, especially in areas where wood is scarce. At the domestic market level, the Petite Côte region possesses the largest number of artisanal fish smoking centres such as Mbour and Joal. About 400 women accounting for 96 percent of all artisanal fish processors are involved in this activity.

Senegal is a major fish drying centre for West and Central African countries. Generally, fish is dried on mats, cement floors, old fishing nets and sometimes directly on the ground.

As in the case of Ghana, inshore fishing vessels usually carry ice onboard to preserve their catch. Some canoe fishermen, who use hook and line for high value demersal species, also carry ice to sea in insulated containers kept in special compartments in their canoes. Icing certain fish species after capture can be logistically difficult and uneconomic for artisanal fishermen. Pelagic species such as sardinella and mackerels do not command high prices and therefore will not attract higher prices when kept in ice.

There are a number of traditionally processed fish products in Senegal. The most common ones are locally called *Guedj* (fermented and dried fish), *Kethiak* (roasted, salted and dried cardinals and Ethmalosa), *Yosso* (smoked juveniles fish), *Sali* (salted and dried fish), *Tambadiang* (small fish salted and dried fermented or not, in bulk), *Yeti* (fermented salted Murex) and *Yoss* (dried juvenile fish).

Fish marketing

As in the case of Ghana, during the major fishing season when supply increases, prices are lower than during the lean season when supply is lower than demand. Fish prices also increase as distances from the landing sites and catch areas increase. Prices for salted and dried fish paid by semi-wholesalers at the Kafountine market in Senegal in 2001 ranged from 300 to 550 CFAF per kg of fish depending on the product. The price of smoked bonga paid by semi-wholesalers at the same market in 2001 was CFAF 150 per kg of fish. The semi-wholesalers sell the fish to retailers, who then sell it to the final consumer. When smoked fish products are destined to foreign countries, prices are set according to the size and quantity of the traditional baskets made with fresh palm tree leaves. The quantity varies from 30 to 65kg. Smoked fish products are also exported in plastic sacks, which weigh up to 100 kg. Table 10 shows the average fresh fish prices paid at the Dakar Central Fish Market. The traditional fish market in Senegal is dominated by smoked and dried fish products destined for inland markets and neighbouring countries. Fish is transported with trucks via interstate roads and in some instances with canoes or small passenger boats by water.

Table 10: Fish prices at Dakar Central Fish Market, Senegal, 2001

Species	Average price (in CFAF
Round sardinella	100.0
Flat sardinella	75
Horse mackerel	175.9
Grunt	337.3
Pandora	264.4
Sea catfish	190.0
Seabream	297.9
Jack mackerel	122.0

Source: Dakar Central Fish Market, 2001

Senegal is a major supplier of both frozen and cured fish products to West and Central African countries. The main export markets for Senegal's cured fish products, particularly for sardinella, are Kayes, Mali and Nzérékoré, Guinea. Senegalese women processors in and around Ziguinchor, Senegal, market their own products by accompanying them as far as Kayes and during the peak period even as far as Bamako, Mali. In other instances, traders from Mali travel to Casamance, Senegal, to buy fish, which they transport via Dakar to Bamako by motorised canoe. The trade with Nzérékoré, Guinea, is dominated by Guinean men, who procure the fish at the processing centres in Senegal and transport it by road to Guinea.

Women's contribution to fish trade

In Senegal, 97 percent of artisanal fish processing and marketing is carried out by women. They operate in small markets located in villages as well as in big central markets such as the central market of Dakar, M'Bour and the Kafountine fish market. Women fish processors and traders are organized in local associations, some of whom are federated in regions such as the Association Santa Yalla in the Casamance region, the Women Association of Mbelling and the Women Association of Mbour.

STRATEGY FOR OVERCOMING CONSTRAINTS TO A BETTER INTEGRATION OF SMALL-SCALE FISHERIES INTO FISH TRADE IN GHANA AND SENEGAL

Constraints to the development of small-scale fisheries

The major constraints, which hamper the development of small-scale aquaculture in Ghana and Senegal, are weak extension services, inadequate supply of good quality fingerlings, lack of knowledge of fish pond management and the high cost of pond construction.

Artisanal fisheries development and efficient distribution and trade of fish is constrained by a lack of infrastructure for fish handling, storage, transport facilities, interstate roads and border posts and controls. There is also lack of information in fishing communities as they live in small isolated settlements along the shores of coastal and inland waters and are rarely visited by extension officers. This situation coupled with high illiteracy, hampers their access to appropriate technologies for fish preservation as well as access to information on resources and markets.

Political and civil unrest including military coups in the countries of the West African sub-region also affect fish trade. In most cases, when a military junta or a rebel group overthrows a government, existing trade arrangements are negatively affected.

The economic situation in many countries of the region is alarming and not conducive to the development of fisheries and fish trade. The introduction of Structural Adjustment Programmes (SAPs) in Ghana in the 1980s, for example, adversely affected all sectors of the economy including the artisanal fisheries sector. As a result of these programmes, the purchasing power, especially of the rural population, is still very low. The lack of foreign currency to import fishery inputs such as petrol, fishing gears, outboards engines and canoes resulted in reduced fish landings, which caused increases of fish prices.

All countries that make up the Economic Community of West African States (ECOWAS), have their own fishery policies, many of which are ineffective either due to bureaucracy and lack of willpower on the part of those responsible. An example is the ECOWAS initiative on streamlining fish trade policy in West Africa. The initiative started in the 1970s and no concrete results have yet emerged.

Tariff and non-tariff barriers to regional fish trade pose further constraints to the development of fisheries and fish trade in the region. Malpractices at check points at the borders between States are obstacles to trade in the region. Also, traders and transporters of fish are mostly illiterate and not able to fulfil the requirements of customs, health and police officers.

As far as taxes and import duties are concerned, efforts were made by a number of regional and sub-regional organisations to identify a common trade policy for their respective member countries. So far, very little has been achieved. Member countries failed to enforce regional trade initiative rules. Another effort by ECOWAS to streamline fish trade policy in West Africa was the introduction of a common duty and tax schedule in 1979, but Member States failed to incorporate these special provisions, preferring to apply their own regulations. Under the stipulated common policy, non–tariff barriers were to be removed within four years from May 1981.

The West African Economic and Monetary Union (UEMOA) and ECOWAS had decided that import duties were to be reduced and to be eliminated or harmonised in conformity with the provisions of the treaty. Unfortunately, economic pressures in member states made it impossible to put the new policy into effect.

Improvements needed for better market integration

As mentioned above, wide-ranging improvements in the infrastructure to be used by the artisanal fisheries sector are vital for the improvement of fish trade in West Africa including onboard handling and preservation facilities, infrastructure at fish landing sites, roads, communication and fish processing and marketing infrastructure.

Fish inspection services are vital prerequisites for the production of good quality and safe fish products for trade. A trade information network on sustainable management of fishery resources should be established and information on the utilization of appropriate technologies for smoking and drying fish should be disseminated.

The statistical database on West African artisanal fisheries is weak and should be improved. Awareness, education, training and information on the socio-economic importance of the artisanal fisheries sector will help improve and strengthen the livelihoods of those involved in the small-scale fisheries sector. Regional cooperation and integration offer tremendous opportunities for economic growth and would enable the countries involved to participate in and benefit from a bigger market and provide opportunities for the integration of African economies into the global economy. In the same context, there is also a need for a monetary co-operation programme, which should involve collective measures to set up a harmonised monetary system and common management institutions.

The artisanal fisheries sector is an important source of foreign exchange, employment and animal protein in the region. It has to be strengthened and interregional trade has to be stimulated and increased by creating awareness of market opportunities in the region and developing trade channels and services to support market integration. To achieve these objectives, the following measures may be considered:

- Market surveys should be undertaken in support of artisanal fish products trade in the region to identify priority needs and to formulate an appropriate programme of assistance to improve regional trade in fish and fish products.
- Buyer and seller meetings and seminars may be organised regularly on trade opportunities in the region to bring decision-makers together and help create and increase business contacts.
- Regional information networks need to be strengthened to provide data on raw materials, annual supply and demand as well as consumption patterns. Until recently, INFOPÊCHE and the West African Development for Artisanal Fisheries (WADAF) programme have jointly published *Bonga Flash*, a bi-monthly magazine on fish trade and price trends in the region. The magazine ceased activity due to lack of funds. The publication should be resumed to provide crucial information on fisheries and fish trade in West Africa.
- A directory of exporters and importers of fish and fish products from the region should be published based on data collected from different regional main markets surveys and buyer and seller meetings.
- Liberal trade policies already formulated such as the UEMOA TEC (Reform of Tariff and Tax Policies) and the ECOWAS trade programme should be implemented.
- Rules and regulations on packaging and labelling of goods should be harmonized. Losses of cured fish commonly occur on a considerable scale during transportation because of poor handling, packaging and mislabelling.

- Smoked fish product standards, quality assurance and certification standards and requirements should be developed in accordance with international requirements.
- The role of existing trade information networks such as INFOPÊCHE should be expanded to establish contacts and promote information flows among artisanal fishmongers such as the fish mammies in Ghana, to create marketing bodies in different regions and to link up with other sub-regional institutions in Africa.
- Training programmes should be organized to improve the skills of national trade promotion officials by utilising existing institutions available in the region.

6

Small-scale Fisheries and Fish Trade in Southern and East Africa

PRESENT AND FUTURE MARKETS FOR FISH AND FISH PRODUCTS FROM SMALL-SCALE FISHERIES IN MOZAMBIQUE

Fisheries and Aquaculture Production and Resources

Resources

Mozambique is endowed with rich fishery resources both marine and freshwater.

The country has a 2 750 km long coastline bordering Tanzania in the north and South Africa in the south. The marine waters cover an area of about 100 000 km^2 within the exclusive economic zone (EEZ). Inland waters cover an area of about 13 000 km^2.

The marine fishery resources are mostly located in the two major continental shelf areas, i.e. the Sofala Bank in the central part of Mozambique and the Delagoa Bank in the south. The main fishing areas are located at the Sofala Bank, Inhambane, Vilankulo, Chiluane and Beira.

The most important marine species include deep water crustaceans such as prawns, lobsters and crabs, shallow water shrimps, pelagic and demersal fishes, shellfish and marine algae and seaweed.

The freshwater fisheries are located in Lake Malawi/Niasssa and Lake Cahora Bassa. Freshwater fishes include tilapia and small pelagics locally known as *kapenta*.

Table 1: State of fishery resources in Mozambique

Stocks exploited or fisheries managed	Level of exploitation
Crustaceans	
Shallow water shrimp	Intense
Deep-water shrimp	Moderate
Line fish	Intense
Deep water lobster	Intense
Rock lobster	Low
Deep water crab	Moderate
Marine fish	
Large demersals	
Large pelagics	Very low
Small demersals	Low
Small pelagics	Low
Molluscs and other resources	
Squid	Moderate
Sea cucumber	Intense
Cephalopods	Low
Freshwater fish	
Kapenta (in Cahora Bassa)	Intense
Demersal fish (in Cahora Bassa)	Intense
Utaka (in Lake Niassa)	Low

Source: FAO, 2005

According to the Ministry of Fisheries in Mozambique, the most valuable stocks of prawn and demersal fish targeted by the industrial and semi-industrial fishery sectors have been assessed to be highly or fully exploited while other fishery resources including large and small pelagics are only lightly exploited or underutilized in remote areas along the coast.

Excellent conditions exist in the country for the development of aquaculture particularly for the farming of prawns, oysters, mussels, algae and pearls. Mozambique has a considerable number of native marine species suitable for farming, the prominent ones being *Penaeid* prawns, i.e. black tiger prawn, Indian white prawn, pink prawn and *kuruma* prawn. Increased farming of prawns could reduce the pressure on stocks presently exploited by capture fisheries. Increased production of freshwater fish could also provide more fish for the local population and reduce the deficiency of the fish supply in the country.

Production

As shown in Table 2, the annual fish production of Mozambique was estimated at 105 000 tonnes in 2004 with the largest production segments obtained from the marine shallow water shrimp and the deep water shrimp fisheries. Over the years, the shrimp fisheries have been the economically most important fisheries in terms of national revenue earnings. The shrimp fisheries account for about 40 percent of the total exports. The freshwater small pelagic fishery for *kapenta* in Lake Cahora Bassa's Barragem is the third most important fishery in terms of exports. Freshwater fisheries for tilapia, catfish and carps are important for the small-scale fisheries sector.

The Mozambican fisheries can be divided into industrial, semi-industrial and small-scale fisheries. The industrial and semi-industrial sectors accounted for about 43 percent of the recorded catch in 2004 while the artisanal sector accounted for about 57 percent of total fish catches. More than 90 000 people are involved in the fisheries sector in fishing, gathering of aquatic organisms, processing and marketing. Seventy-thousand people are involved in the marine fisheries sector and 20 000 in freshwater fisheries. About 500 000 people depend indirectly on fishing activities for their livelihood. Thus, fisheries in Mozambique are of particular importance for the national economy as they provide employment and a source of animal protein and foreign currency earnings.

In 2004, fish caught by the industrial and semi-industrial fleets amounted to almost 30 000 tonnes, of which 65 percent were exported, resulting in net foreign earnings of about US$94 million. The fisheries sector contributed about 4 percent to the gross domestic product (GDP) and accounted for about 10 percent of the total export earnings of Mozambique. Government revenues from licence fees and levies amounted to US$3.8 million in 2002.

Table 2: Total fisheries production of Mozambique, 2000-2004 (in tonnes)

Sector/year	2000	2001	2002	2003	2004
Industrial and semi-	25 915	20 501	22 185	23 112	29 878
Artisanal	63	55	47	67	60
Tuna fisheries	5 081	3 096	3 000	7 450	14
Total	94	78	72	97	105

Source: Ministry of Fisheries. Maputo, Mozambique, 2005

Marine Fisheries

Marine fisheries account for almost 90 percent of Mozambique's total fish production. The main marine fishery resources include crustaceans such as prawns, deep water shrimp, crayfish, lobsters, crabs, marine finfish, both demersal and pelagic species such as groupers, snappers, emperors and sea breams including highly migratory tuna species like yellowfin, bigeye and albacore tuna, swordfish, sharks, cephalopods, molluscs, squid, octopus, sea cucumbers and bivalves.

Inland Fisheries

Inland water bodies include Lake Niassa, the third largest lake in Africa and third deepest worldwide, the man-made Cahora Bassa Lake and a great number of rivers. Lake Cahora Bassa and the Mozambican part of Lake Malawi provide fisheries related livelihoods for about 20 000 people. A total of 10 000 tonnes of small pelagics, i.e. *kapenta* and demersal fish from Lake Cahora Bassa are caught, processed and marketed each year, of which 4 000 tonnes are caught by artisanal and small-scale fishers.

Aquaculture

The development of aquaculture of shrimp, bivalves, molluscs and tilapia is being actively promoted by the government. Both freshwater and marine aquaculture benefits from the diversity of the natural environment and the availability of suitable native species for farming. While freshwater culture is mainly focusing on integrated fish farming systems to improve the population's diet, marine aquaculture is oriented towards both low cost protein supplies and high value products for exports. Particular attention is given to marine prawn culture because of its economic importance.

Aquaculture production is still very modest in terms of volume. In 2002, the total aquaculture production of Mozambique was 855 tonnes representing a value of about US$3 million, which included 600 tonnes of marine shrimp, 100 tonnes of freshwater fish and 155 tonnes of marine algae as shown in Table 3. According to the Ministry of Fisheries in Mozambique, 900 tonnes of tilapia species are produced annually at a small-scale level for subsistence purposes. Freshwater aquaculture started in the 1950s at an artisanal level while marine aquaculture started in 1995 at an industrial level. The main marine species farmed in Mozambique include black tiger prawn (*Penaeus monodon*), Indian white prawn (*P. indicus*), pink prawn (*Macrobrachium monoceros*), *kuruma* prawn (*Modiolus philippinarum*), bivalves (*Perna perna, Meretrix meretris, Modiolous philippinarum, Eumarcia pauperculata, Sacrostrea cucullata, Cassostrea gigas,* and *Veneruspis Japonica*) and mud crab (*Scylla serrata*).

Farming of black tiger and white prawn is oriented towards export. In 2003, the production of cultured shrimp from a 132-ha farm was reported to be 300 tonnes. Commercial shrimp farms are presently situated in Beira (a 132-ha farm), Quelimane (300 and 150-ha farms) and Pemba (a 250-ha farm). Total employment in the aquaculture industry is estimated at about 500 to 600 people. Freshwater aquaculture is dominated by the farming of native species. The popular species farmed in Mozambique include tilapia (*Oreochromis niloticus*), African catfish (*Clarias mossambicus*), freshwater shrimp (*Macrobrachium rosenbergii*) and carps (*Cyprinus carpio*).

Tilapia farming is currently carried out as small-scale cage culture with a production potential of about 80 tonnes per year. There are about 2 750 subsistence farms with an average size of about 100 m^2 oriented towards family consumption. Other important aquaculture species are seaweed, oysters and crabs.

Seaweed (*Kappaphycus* spp.) farming involves about 5 400 people, of which 65 percent are women. The production of seaweed in 2003 was 523 tonnes.

Table 3: Aquaculture production of Mozambique, 2001-2002 (quantity in tonnes, value in US$1 000)

Year	2001		2002	
	Quantity	Value	Quantity	Value
Aquaculture production	70	21	855	3 033

Source: Ministry of Fisheries, Mozambique

The Artisanal, Semi-industrial and Industrial Fisheries Sectors

The artisanal fisheries sector

Artisanal fishing is carried out along the entire coastline of Mozambique. It is of particular importance in the provinces of Nampula, Zambézia, Sofala, Inhambane and Maputo. Artisanal marine fisheries are a source of income for about 50 000 families. The artisanal fisheries sector accounts for about 90 percent of the 90 000 people employed in the fishery sector. The artisanal fishing fleet consists of 15 000 artisanal fishing boats and canoes. Sixty percent of the total catch in Mozambique is contributed by artisanal fisheries. This includes artisanal shallow-water shrimp fisheries, which land 5 000 to 6 000 tonnes of shrimp annually.

Fisheries at both Lake Malawi and Lake Cahora Bassa are artisanal and supply mainly small pelagic fish species for the domestic market and the regional African market. Small-scale and artisanal fisheries in Mozambique play a significant role in the national economy. They provide employment, income and a source of animal protein for the majority of coastal communities. Artisanal fishing is carried out by individual or small groups of fishermen using mostly non-motorized fishing vessels of 3 to 8 m in length. Only 3 percent of the artisanal fishing boats are motorized. Beach seines, gillnets and longlines are the most common fishing gears used to catch fish. The artisanal fisheries sector also includes fish collectors and divers. Artisanal fishermen and their families have a low socio-economic status. Marine and freshwater fisheries are carried out from 787 fishing centres, of which more than 500 are scattered along the coast. While the industrial and semi-industrial sectors land their catches at designated fishing ports, artisanal fishers land their catches mostly at beach landing sites. The catch composition of artisanal fisheries in Mozambique is shown in Table 4.

Table 4: Semi-annual landings by artisanal fisheries in Mozambique by species, 2004-2005

(in tonnes)

Species/year	2004	2005	Percentage change in
Crabs	130	88	-32 %
Fish	27 776	29 756	+7 %
Shrimp	1 465	2 718	+86 %
Cephalopods	107	162	+51%
Sharks	184	175	-5%
Others	2 853	1 139	-60 %
Total	32 515	34 038	+5 %

Source: Ministry of Fisheries, Maputo, Mozambique

In view of the importance of the artisanal sector, the government places great emphasis on the further development of this sector and has established the Institute for the Development of Small-scale Fisheries (IDPPE) for this purpose.

The Industrial and Semi-industrial Fisheries Sector

The industrial fisheries sector in Mozambique consists largely of joint ventures between the government and foreign fishing companies from Japan and Spain. Seventy percent of the Total Allowable Catch (TAC) of Mozambique for shrimp has been allocated to these joint ventures. The shrimp industry based in Beira and Quelimane is export-oriented and represents an important source of foreign exchange income for the country. Seventy percent of the production comes from two major commercial companies, i.e. PESCAMAR and EFRIPEL. The catch is frozen onboard the fishing vessels and exported directly to

Japan and the European Union. Well-equipped foreign shrimp fishing fleets are active in Mozambique's waters. About 187 national industrial fishing vessels are operating also in the industrial fisheries sector. The main commercial species targeted by the industrial sector include lobster, crabs, deep water shrimp, fish, shallow water shrimp, crayfish and squid.

The semi-industrial fleet consists of about 213 vessels, of which 116 are targeting small pelagics, i.e. kapenta.

Main Species Exploited by the Various Fisheries Sectors in Mozambique

The marine species exploited in Mozambique's waters include both demersal and small pelagic species such as groupers, snappers, emperor, seabass, croakers, dentex, pompano, parrotfish, flat needlefish, seabream, sardines, dorado, Spanish mackerel, shallow water shrimps, horse mackerel, barracuda, chub mackerel, bivalves, sea cucumber, crabs, cephalopods, octopus, squid, oysters and clams. The freshwater species exploited in Mozambique are tilapia, catfish and carps.

The main marine species harvested by small-scale fishers are species found in the shallow waters of the continental shelf. Small pelagics, sardines and crabs are caught by purse seines, gillnets and beach seines while large fish such as pompano, groupers, snappers and seabream are caught with longlines. The pelagic fishes, mainly sardines and horse mackerels, dominate the catch composition and account for about 70 percent of the production.

Demersal and pelagic fishes are the principal finfish species targeted by the small-scale and artisanal fisheries. Fishes such as groupers, snappers, dentex and croakers are mostly targeted because they are bought by the industrial processing companies. A part of the catch is consumed in the domestic market. Once fish is landed on the beach, the catch is sorted according to sizes. The fishes are then divided into groups based on their size and commercial value.

The shrimp fishery has the highest commercial value and is the only fishery for which a Total Allowable Catch (TAC) limit has been set. Shrimp is harvested in deep waters by the industrial fleet and most of the catches are frozen onboard for export. The artisanal and small-scale fishing fleet targets shallow water shrimp and lands 5 000 to 6 000 tonnes of shrimp annually. Most of the catch from the small-scale fisheries sector is frozen or sold fresh and a certain part is dried. In the domestic market, fresh shrimp are mainly sold in big cities like Maputo, where customers with high purchasing power are located. Most of the catch, however, is used by commercial fish processing companies for export to international markets.

Crustaceans, bivalves, cephalopods and shellfish are harvested in small quantities. Squid and crab are of high commercial value. Because they are landed in small quantities, they are only distributed in special and limited markets. Squid is usually purchased by local restaurants and markets in Maputo and Beira and consumed both in fresh and dried form.

Oysters are caught by collectors along the coast and processed by boiling and sun-drying. Oysters have a good market in Vilankulo, Massinga, Morrumbene and Maxixe and are sold in cans and glass jars. Lobsters are caught by divers and consumed fresh. The main markets are Maputo and Beira and consist mainly of local restaurants.

Tilapia, catfish and a small quantity of carp species are among the most common freshwater fish species caught and processed exclusively by artisanal and small-scale fishers; the quantity is very limited though. The products have a ready market in local communities. Most of the catches are consumed in fresh form as the local market prefers fresh fish. Tilapia is also smoked for preservation. Fish from inland waters is caught mainly for subsistence purposes.

Fish Processing

Fish processing in Mozambique traditionally involves a number of different methods often used in combination. Salting and drying of fish is practiced in the northern and central part of the country, while freezing of fish is also used in the northern and fish smoking in the southern part of Mozambique. In many parts of the country, lack of cold storage facilities forces processors to rely on traditional processing methods. Most fish processing takes place onboard industrial and semi-industrial vessels such as the freezing of shrimp. Value-added fish processing on shore is promoted through long-term fishing rights assignments.

At present there are some 14 fish processing and freezing plants that are EU-accredited. Of these, five are located in Maputo, five in Beira and four in the northern part of the country. These plants buy their raw material from both semi-industrial and artisanal fishermen.

Their main product is frozen shrimp but they also process various demersal fish species. In addition, there are some 40 smaller processing plants and operations that are not accredited by the EU. These smaller plants supply the local and regional markets with fresh, frozen, dried and salted fish.

The processing plants often provide ice and plastic boxes to artisanal fishermen in order to ensure a higher quality of raw material, which the plants procure from fishermen, and to oblige the fishermen to deliver their catch to the company, which provides them with ice and boxes.

Sun-drying without the use of salt and salting and drying are the two main processing methods for preserving fish in the small-scale fisheries sector. Salting and drying is used for squid and fish of all sizes and also for fish of second-grade quality. The process involves the following stages: for big-size and medium-size fish, the fish is split open, washed and then salted before it is sun-dried. Small-size fish is not gutted but salted and dried whole. The whole process takes about 12 to 36 hours depending on the size of the fish.

Sun-drying without salting is mostly applied to small fish and small shrimp. Fish are put directly on the ground and, to a lesser extent, on racks, plastic sacks and banana leaves. The process takes about 12 to 24 hours depending on weather conditions. Fish processors and traders try to sell fish while it is fresh, as fresh fish generally fetches a higher price.

Fresh fish brings in 20 to 25 percent more revenue than dried fish locally known as *macakua*. In general, fish processed by artisanal and small-scale processors is contaminated with sand and other external agents. Most of the landing sites lack clean water to clean fish and fish products.

Consumption Patterns and Marketing Channels for Small-scale Fisheries Products

Mozambique's average fish consumption per capita is estimated at 7 kg per year. Fish consumption is higher in coastal areas, where it ranges from 10 to 12 kg per year.

The pattern of fish consumption differs between urban and rural areas and is influenced by cultural beliefs and financial status. People living along the coast prefer fresh fish while the access of those living in inland areas is often limited to dried fish products. Frozen horse mackerels and dried sardines are the staple fish foods for low-income consumers as they are cheap, readily available and have a long shelf-life.

Demand for fish products in Mozambique exceeds what the domestic fishery industry can supply. Because the major part of good quality seafood is exported to generate foreign currency, Mozambique depends on fish imports in the form of frozen small pelagics such as horse mackerels to meet domestic demand. It is estimated that 25 000 to 30 000 tonnes are annually imported from Angola and Namibia. It is expected that the demand for seafood will grow substantially until 2025 and there is a need to increase supplies. Table 4.5 shows supply targets for fish products in Mozambique set by the Ministry of Fisheries.

Table 5: Supply targets of fish products in Mozambique, 2000-2025

Source		Target catch for domestic			(tonnes	
	2000	2005	2010	2015	2020	2025
Industrial and semi-industrial fisheries	33 631	44 421	55 211	66 001	76 791	87 581
Artisanal fisheries	84 065	101 460	118 855	136250	153 645	171 040
Total	117 696	145 881	174 066	202 251	230	258

Source: Ministry of Fisheries, Mozambique, 2005

The small-scale fisheries production is mostly oriented to the domestic market. However, small amounts of shrimp, high quality fish and lobsters are exported through some of the fish processing companies based in Nampula, Sofala, and Cabo Delgado.

The distribution of fish products depends on their commercial value and quality. Most fresh fish from the small-scale fisheries sector is consumed by people living close to landing sites. Otherwise, fish is either salted and sun-dried or smoked for long distance transportation. The national distribution system for fish, especially fresh, is not well developed.

The distribution chain for fish products is simple and three main distribution channels can be distinguished.

(1) *Fishermen – Consumers*. This is the shortest channel in the distribution of fish. Fresh fish is sold by fishermen directly to consumers as soon as it is landed at the beach.

(2) *Fishermen – merchant retailers – consumers*. This is the typical channel in the informal distribution of fish. Unlike the previous channel, this channel involves a number of informal retailers. The retailers are of both sexes and of different age groups. They buy fish from fishermen immediately after it has been landed to resell it within the vicinity of the landing site. The retailers make use of beach markets, landing points, fishing centres, informal coastal markets and municipal markets to sell their fish. The amount of fish they buy from fishermen is usually very small and rarely exceeds 50 kg. In general, the retailers buy and sell between 15 to 30 kg of fresh fish at a time.

(3) *Fishermen – processors – fish merchants – consumers*. This distribution channel is mostly used for processed fish products, mainly dried and smoked products. Fishermen in the inland areas of Bazaruto, Santa Caroline, Santo Antonio and Santo Isable, for example, process their own catch. The fish is later sold to traders for reselling to consumers in local markets. Farmed fresh fish is also distributed in this way.

Fresh products are distributed in local markets of big cities. The products are kept on ice or in cold water at local markets. If kept in water, the products do not last long and deteriorate within two days.

Some market centres have freezers, where traders can keep the fish overnight for a fee. The use of ice is very limited as it is only used for high value fish species and products. There is a limited number of ice-making machines in the country. In most cases, fishermen obtain ice from processing companies if they agree to sell their fish to them. If fish is not sold fresh, it is frozen or processed by drying and smoking.

Fish merchants represent the most important link in the distribution chain of fish products in local and regional markets. They can be divided into three groups based on the amount of fish they handle. Large-scale wholesale merchants, locally called *grossistas,* are agents of processing companies and managers of big enterprises that buy large volumes of fish directly from fishermen for redistribution or for processing. Most of these merchants process their fish by freezing or drying and sell it through their own enterprises or in retail markets. They purchase up to 1 000 kg of fish at a time. In addition to supplying the processing factories, these merchants also supply local retail markets, local restaurants and local fishmongers.

Medium-scale fish merchants, locally called *semi-grossistas,* buy both fresh and processed products from fishermen and small processors to resell these in local markets. The *semi-grossistas* purchase between 50 and 200 kg of fresh fish and 1 to 3 sacks of dried fish per trip to the market. They usually make two to three trips per month to collect fish at fish landing sites. Most of them have their own market space in the main market centres of Beira and Maputo.

The third group of fish merchants are merchant retailers, who are involved in petty trade. They purchase very small quantities of fish at a time to resell to street vendors in street markets and in municipal markets located in areas close to their own home. Mostly women and young children are involved in this type of trade.

The distribution of fresh fish to inland markets is hampered by poor roads and by a lack of adequate storage facilities for fresh or frozen fish throughout the country. Processed fish such as dried or salted-dried fish is usually packed in sacks and transported by bicycles or pick-up trucks to local fish markets in villages and in larger towns and cities.

The Domestic Market for Fish and Fish Products

Mozambicans are traditionally fish eaters and prefer fish to red meat unlike consumers in other African countries such as Namibia, South Africa and Botswana, where red meat and chicken are the staple foods in the traditional diet. Most of the local fish markets in Mozambique are concentrated in the more densely populated regions.

There are basically two major consumer groups in the domestic market. The first group is represented by the high-income group of European and Mozambican origin, which is concentrated in the major cities such as Maputo, Beira, Sofala and Manica. This group prefers high value fresh and frozen fish products of good quality. It also comprises hotels and restaurants, which buy much of the crustaceans and cephalopods sold in Mozambique. Most of the fish is obtained directly from fishermen, retail markets and fishmongers.

The second group of domestic consumers consist of low-income consumers both in cities and in rural areas with a traditional diet. Small pelagics, horse mackerels, sardines and juvenile fish both of demersal and pelagic species in fresh and cured form are the fish products preferred by these consumers as these products are relatively cheap. Women street vendors sell small quantities of dried small pelagics and dried shrimp together with other agricultural commodities such as beans and vegetables.

The most common markets for African products are local street markets, where women between 14 to 55 years of age and young men sell all sorts of fish products ranging from small quantities of dried fish to fresh fish. In general, fish trade in Mozambique is slow during the rainy season when vegetables are abundant.

Non-African International Markets

The international market for Mozambique's fish products is diverse and includes Africa (the Republic of the Congo, Malawi, South Africa, Zambia and Zimbabwe), Asia (China, Hong Kong SAR and Japan) and Europe (Italy, Portugal, Spain and the United Kingdom).

The European Union (EU) accounts for about 63 percent of Mozambique's fish exports by volume, Africa for 25 percent and Asia for 12 percent.

High-value prawns are the principal product currently being exported to the EU (mainly Spain) and Japan. The value of fish exports from Mozambique to international markets was US$98.5 million in 2002 and US$87 million in 2003.

Table 6 shows the destination of exports of Mozambique's fish products in 2001. Table 4.7 shows volume and value of foreign trade in fish products in Mozambique for the period 1997–2003.

Table 6: Destination of exports of Mozambique's fish products in 2001

Destination	Quantity (in tonnes)	Percent
AFRICA	3 425	24.9
South Africa	1 782	13.0
Malawi	84	0.6
Congo	206	1.5
Zambia	235	1.7
Zimbabwe	1 118	8.1
ASIA	1 682	12.2
Japan	1 669	12.1
China, Hong Kong SAR	13	0.1
EUROPE	8 638	62.9
Spain	4 737	34.5
Portugal	3 658	26.6
United Kingdom	1	0.0%
Italy	242	1.8%
TOTAL	13 743	100.0

Source: Ministry of Fisheries, Maputo

Table 7: Foreign trade in fish products in Mozambique, 1997-2003[5] (quantity in tonnes, value in US$1 000)

Year	Exports		Imports	
	Quantity	Value	Quantity	Value
1997	10 367	85 400	5 083	5 438
1998	9 463	71 626	14 518	7 866
1999	9 556	76 867	18 740	10 350
2000	11 511	103 716	21 970	8 177
2001	14 191	94 798	20 238	6 980
2002	11 663	98 559	17 835	7 749
2003	10 412	87 042	12 570	9 495

Source: FAO FISHSTAT, 2005

In summary, Mozambique's foreign trade in seafood is characterized by exports of high-value fish products and imports of low-value fish.

African Export Markets

The main African markets for fish and fish products from Mozambique are countries of the Southern Africa Development Community (SADC) and the Southern Africa Customs Union (SACU). The SADC and SACU markets account for about 25 percent of all Mozambican fish exports. As shown in Table 6, South Africa and Zimbabwe reported the highest import volumes of fish products from Mozambique in 2001. Fish processed onboard semi-industrial Mozambican fishing vessels is mainly exported to South Africa. Freshwater small pelagics (*kapenta*) caught by the semi-industrial fisheries sector are exported in large volume to Zimbabwe.

Traditionally processed, i.e. salted and dried and smoked fish is exported informally to neighbouring countries such as Tanzania, Malawi, Zimbabwe, the Republic of the Congo and Zambia. The informal border trade is done by petty traders, who transport small quantities of fish at a time, mainly to avoid taxation. There are consequently no records of this trade and it is very difficult to estimate the volume traded. Fish exports to Tanzania originate from the Cabo Delgado Province. The export to Malawi originates from Nampula, Zambezia and Beira. Fish exported to Zambia originates from Tete at the Cahora Bassa Lake. Most of the products are exported in dried and smoked form with the exception of South Africa, which imports frozen shrimps and fish.

Intraregional trade in the SADC region is currently underdeveloped. In 2001, only about 150 000 tonnes of fish products were distributed within the region. This represented only 10 percent of the 1.5 million tonnes of fish produced annually in the SADC region. Intraregional fish trade is suffering from the meat-based diet of many African countries. The lack of knowledge about fish preparation and traditional preferences for meat are hindering market penetration and expansion.

In addition, barriers to intraregional fish trade are represented by inadequate infrastructure for large trade volumes, lack of transport, storage and distribution facilities and high import taxes. Furthermore, foreign exchange is lacking and export credit facilities are poorly developed. Unrecorded cross-border trade of low volumes of fish to avoid taxation is very common. This informal intraregional trade needs to be studied through data collection and analysis to determine its impact on national economies.

Involvement of Women in Fisheries in Mozambique

In spite of fishing and fisheries being seen as a basically masculine activity, the involvement of women is on the increase. Although the exact number of women involved in fisheries related activities is not known, it is estimated that of the 100 000 people working in the artisanal fisheries sector of Mozambique, 20 percent are women. In Mozambique, as in most African countries, women are mostly involved in fish processing, collection and trading while men are involved in fishing operations. Twenty-five percent of women involved in fisheries related activities are fish traders.

In some cases though, women are also participating in capture fisheries together with men. Women participate in fishing activities in coastal areas and estuaries particularly in fishing for small pelagics and small shrimp and in beach seine fishing together with fishermen. In 2005, South Africa reported that a number of women were fishing at sea as crew on trawlers. In Mozambique, it has also been observed that women have started to participate in fishing operations at sea.

Apart from fishing, women participate in other forms of harvesting of aquatic organisms. It is estimated that about 35 percent of women in Mozambique participate in the collection of shellfish, i.e. oysters and mussels. The collection of shellfish is a predominantly female activity. According to the Institute for Small-scale Fisheries Development (IDPPE), women collectors represent about 71 percent of the 23 500 registered shellfish collectors in the coastal zone. The collection of shellfish is mostly a subsistence activity for supporting the family and it involves all age groups.

There are also women, who collect molluscs for commercial purposes. Molluscs such as clams have a ready market in the south while oysters are sold in the northern part of the country. The most important problems confronting women collectors are the lack of ice and cold storage, lack of transportation for taking their products to markets and the difficulty of obtaining credit to improve their activities.

In many cases, women participate in fisheries activities as partner and support to their husbands rather than as the sole owner of the business. Commonly, women wait at landing sites for their husbands to land the catch. They then sort the fish by species and process or sell it fresh. Women, who are not involved in family based activities together with their husbands, are usually organized in groups and associations of women processors and the responsibilities are shared among the group members. Many of these women are divorced, widowed or single mothers struggling to support their families.

Women are frequently involved in selling of fresh fish, shrimp and molluscs rather than in the selling of processed fish. The latter activity is dominated by men. Fresh fish is usually marketed close to landing sites and in the main cities. Some women from fishing communities trade fish in the vicinity of their own home and move around from door to door carrying their products in baskets on their heads. Some of these women travel as far as 50 to 80 km from the landing site to the closest fish markets in the city. Normally, it takes two to three weeks to return and procure the next supply of fish. The women use their own traditional means to preserve the fresh fish from deteriorating during the transportation process.

The aquaculture industry is another sector where women play an important role. In Mozambique, the aquaculture of algae and freshwater fish are some of the activities, in which a large number of women are involved. According to the IDPPE, about 3 600 people were actively involved in aquaculture in 2004, of whom some 70 percent were women. Many seaweed farms are owned by women, who are themselves directly involved in selling and promoting their products. Women are specifically involved in seaweed farming for the purpose of generating income. With regard to freshwater fish farming, women are in most cases involved together with their husbands at a small-scale subsistence level and the production is basically used for the purpose of feeding the family.

The involvement of women in fisheries in Mozambique is very important as women serve as a bridge between producers and consumers in the processing and distribution of fish and fish products. Although women play such a crucial role in fisheries related activities, they still do not receive enough support from public and private institutions. Women traditionally are required to take care of the family while men go fishing. This practice has contributed negatively to the recognition and support of women in the fisheries sector. Women are often marginalized and ignored when it comes to financial support for their fisheries related activities. Women are also often discriminated against when it comes to credit facilities and money lending even though they usually have a higher savings rate and are generally more responsible than men.

Improving Marketing Efficiency for Fish Products from Small-scale Fisheries

A large portion of fresh and processed fish products from the small-scale fisheries sector caters to the domestic as well as to regional markets. Traditional methods of processing, i.e. sun-drying and smoking are not entirely appropriate for handling the large volumes of fish now moved commercially. Significant post-harvest losses have been incurred due to the deterioration of the quality of fish and fish products at fish markets and during transportation. The improvement of product quality would increase the value of the product and generate new income opportunities.

The introduction of improved and more advanced handling and processing methods, including the use of ice at landing sites and marketing areas, and new designs for drying facilities and smoking kilns, could improve the current products of the small-scale fisheries sector and provide high quality products that could generate better returns. While the fish products for exports are meeting high quality standards, the products for the local and regional markets are not subject to any quality assurance procedures.

Target markets for improved small-scale fisheries products could be both local and regional markets. Mozambique, being a member of SADC, can take the advantage of free trade agreements to increase intraregional fish trade with other SADC member countries. A more direct link in the marketing chain should be generated. For instance, direct collaboration between fishermen and processing plants through the provision of cold storage facilities would reduce post-harvest losses and have a positive impact on distribution of fish in both local and international markets.

CONSTRAINTS TO AND NEEDS FOR THE DEVELOPMENT OF SMALL-SCALE FISHERIES IN MOZAMBIQUE

Constraints and Problems

The main obstacles to the development of the small-scale fisheries sector and its better integration into fish trade in Mozambique are a lack of knowledge, poor technology and equipment and difficulties in obtaining credit. There is also lack of information related to fish prices and markets. Unhygienic fish handling, preservation and processing methods used in the small-scale fisheries sector in Mozambique result in large post-harvest losses, low quality of fish products in local and regional markets and difficulties in the placement of fish products with high commercial value in competitive markets.

The problems outlined above are mainly caused by poor conditions at fish landing sites, lack of ice and cold storage at landing cites and onboard fishing boats and lack of knowledge on proper fish handling procedures by fishermen. The infrastructure for fish distribution to potential markets is unsatisfactory. The involvement of intermediaries such as marketing agents reduces the share to producers of the prices paid by the final consumers.

Needs

Improvements in product quality and marketing efficiency for small-scale fisheries products will require financial investments. IDDPE has been actively involved in the development and improvement of conditions for small-scale and artisanal fisheries in Mozambique. According to the IDDPE, the sector needs more assistance in the form of training, research and infrastructure.

More emphasis should be given to the following needs:

- support to market development initiatives including the establishment of shore infrastructure, landing facilities, auction halls, ice making and cold storage facilities and specialized and well-equipped fish markets in remote areas;
- training in improved methods of fish processing, packaging and storage to ensure better product quality;
- improvement of preservation methods and product quality control to reduce post-harvest losses;
- development of value-added products through onshore processing of products for export;
- proper marketing services for fishermen and women, information on fish prices, alternative trade opportunities and more information on market access;
- improvement of distribution and logistics in the marketing of fish products with high commercial value for competitive markets;
- improvement of quality assurance systems and compliance with national, regional and international market requirements preceded by the establishment of quality standards for fish products that have a high commercial value both in regional and local markets; development of rules of sanitary, hygiene and quality standards and methods for fish products destined for local consumption and markets;

- access for fish processors, fishermen and fish traders to credit and financial services to facilitate investment in essential facilities and services.

PRESENT AND FUTURE MARKETS FOR FISH AND FISH PRODUCTS FROM SMALL-SCALE FISHERIES IN THE UNITED REPUBLIC OF TANZANIA

The small-scale and artisanal fisheries sector is by far the most important fisheries sector in Tanzania. The sector is characterized by small-scale fishermen operating inexpensive boats and fishing equipment. The fish caught is typically processed and marketed by women. People involved in fisheries activities are those who live in coastal communities and along the shores of lakes. Fishermen are as young as 14 years. This includes school drop-outs, who see fishing as an easy way of making quick money.

Tanzanian fisheries are important for its economy because the sector supports a large number of artisanal fishermen, who depend on fishing for their livelihood. In addition, fisheries are a vital source of food, income and employment for the local communities along the 1 450 km long stretch of the Tanzanian coastline, numerous islands and various lakes.

In recent years, Tanzania had average annual fish landings of over 290 000 tonnes while the estimated production potential is as high as 730 000 tonnes. Marine fisheries account for up to 20 percent of Tanzania's annual total landings and 80 percent of the production comes from freshwater fisheries. Small-scale and artisanal fisheries account for about 95 percent of the total catch. Fisheries provide direct and indirect employment to about 0.5 million Tanzanians.

While most of the fish catches from marine and inland waters are used for subsistence purposes, some are exported to African and non-African markets.

Fish products, mainly Nile perch fillets, shellfish, shrimps, lobsters, crabs and octopus are important export products of Tanzania, which account for 10 percent of the nation's foreign exchange earnings. The fisheries sector contributes about 3 percent to the country's GDP.

Fisheries and Aquaculture Production and Resources

Tanzania has a coastline of 1 450 km and a narrow continental shelf. The Tanzanian EEZ in the Indian Ocean has a size of about 223 000 km^2 and a production potential of some 730 000 tonnes of fish per year. The marine fisheries are composed of small-scale fisheries carried out by the people living in coastal areas, while the industrial fisheries are based around Dar es Salaam. Until recently, only one foreign and three local fishing vessels were licensed to fish offshore in the country's EEZ. It is estimated that over 20 000 small-scale fishermen are engaged in artisanal fisheries along the coast of the mainland alone. Another 18 600 small-scale fishermen are operating from the island of Zanzibar. Thus the main share of marine catches is landed by small-scale fishermen using traditional fishing vessels including small boats, dhows, canoes, outrigger canoes and dinghies with and some without engines.

More than 500 species of fish are utilized as food fishes. Reef fishes are the most important species. The major commercial marine species include emperors, snappers, sweetlips, parrotfish, surgeonfish, rabbitfish, grouper, goatfish, kingfish, sharks, rays, shrimp, lobster, sardines and sea cucumbers. Most of these fishes are used for subsistence purposes while some are exported.

Industrial fisheries account for about 5 percent of the total marine catch of Tanzania. The industry comprises three vessels that trawl for shrimp and use seine nets for sardines within the Tanzanian EEZ. The industrial sector is export oriented. Exports of marine products from Tanzania also comprise tuna, shells, lobsters, crabs, squids, octopus and aquarium fish.

Inland Fisheries

Tanzania is well endowed with freshwater fishery resources, which are exploited by artisanal and subsistence fishermen. Inland waters cover about 6.5 percent of the total land area of the country.

The main inland fish species are Nile tilapia, Nile perch, small freshwater pelagics (*dagaa*), catfish and *cyprinids*. Lake Victoria is the second largest freshwater body in the world and the most important lake for Tanzania's inland fish production. Nile perch, which was introduced in the lake in the 1950s, is by far the most economically important species followed by endemic species, i.e. *dagaa* and Nile tilapia. Figure 4.4 shows the freshwater fish production from 1986 to 2003.

Aquaculture

Tanzania has a good potential for aquaculture considering the large coastal belt, which could support brackishwater fish farming, and the existence of hundreds of water storage reservoirs and other small water bodies.

Fish farming started in Tanzania between the early 1940s and the late 1960s. The main species farmed are Nile tilapia (*Oreochromis niloticus*) and catfish (*Clarias gariepinus*) on a small scale and rainbow trout on a medium scale. Mariculture mainly involves seaweed farming including the farming of *Kappaphycus cottonii* and *Eucheuma spinosum*. Farming of prawn is still at an infant stage and the main species farmed is black tiger prawn (*Penaeus monodon*). Fresh fish and prawns are cultured in ponds while seaweeds are farmed in the open sea within the intertidal zone. Freshwater fish farming is carried out along rivers in the Morogoro, Arusha, Ruvuma and Mbeya regions, where farmers can divert river water to their ponds. Prawn farming is done in the Pwani region, particularly in Mafia Island. It is estimated that at least 14 000 people are involved in aquaculture activities. Tanzania does not allow aquaculture activities in Lake Victoria because of the fear that aquaculture operations might contaminate the lake. Table 8 shows that aquaculture production in Tanzania has exceeded an annual level of 7 000 tonnes in 1999 and remained stable.

Table 9: Aquaculture production of Tanzania, 1994-2003 (in tonnes)

	1994	1995	1996	1997	1998	1999	2000	2001	2002	2003
Eucheuma seaweeds	3 000	4 000	3 000	3 000	5 000	7 000	7 000	7 000	7 000	7 000
Nile tilapia	150	200	200	200	200	200	210	300	630	2
Total	3 150	4 200	3 200	3 200	5 200	7 200	7 210	7 300	7 630	7 002

Small-scale Fisheries

In Tanzania, the marine small-scale fisheries sector is by far the most important sector of the marine fishery industry. Fish is landed by small-scale fishing boats at many small landing sites.

In mainland Tanzania, 95 percent of the total marine catch is landed by small-scale fishers. In the island of Zanzibar, marine fisheries are exclusively artisanal. The landings consist of multiple species. Reef fishes alone such as emperors, snappers and groupers (Serranidae) make up one-third of the catches. Approximately one-third of the total marine catches comprise small pelagic fishes such as sardines (Clupeidae), anchovies (Engraulidae), small mackerels (Scombridae) and horse mackerels. Large pelagic fish species include jacks and trevallys (Carangidae), kingfish (Scomberocoridae), tunas, mullets and swordfish. Other important species are sharks and rays, crustaceans (shrimps, lobster and crabs), octopus, sea cucumber, gastropods, bivalves and shellfish.

In inland fisheries, all fish is landed by artisanal fishers. The landings consist of Nile perch, small pelagics (*R. argentea*) and Nile tilapia. The use of ice is very limited. In the case of Nile perch, the fish is iced only in the trucks, which transport the fish after being sold to agents or representatives of the processing factories. Efforts are under way to improve hygiene and quality of fish onboard during fishing and fish landing. Fish inspectors are recording landings and carry out sample controls.

With regard to fish farming, the private sector is working together with small-scale farmers. The farmers produce seaweed and finfish for family consumption and to a small extent fish for local markets. The government provides technical support and fingerlings to farmers.

Long-term Production Trends

Fish production declined sharply to 295 200 tonnes in 1994. Fish production picked up again in 1995 and stabilized since 1997 around a level of 350 000 tonnes per year. The production trends of marine and inland fisheries in Tanzania are shown in Tables 9 and 10. Tables 11 and 12 show marine fisheries landings in Zanzibar.

The relatively stable marine fisheries production in the mainland as well as in Zanzibar from 1997 onwards as shown in Tables 9 and 10 may indicate that the marine fishery inshore resources accessible to small-scale fisheries have reached their maximum level of exploitation. Artisanal fishermen can only increase their production if they acquire larger motorized fishing vessels, which enable them to fish in deep waters. While the quantity of the marine production remained at a similar level, the value of the production increased. As also shown in Table 4.10, the production from inland fisheries slightly declined while the value of the freshwater fish production increased over the same period.

Lake Victoria Fisheries

In the 1980s, this fishery became more important and the production reached 46 800 tonnes in 1984. It reached its peak in 1990 when nearly 180 000 tonnes were landed. The production then declined to about 100 000 tonnes in 1991 and increased again in 1993, 1995 and 1998. The Nile perch production stabilized between 90 000 and 100 000 tonnes between 1998 and 2003.

Table 9: Fish catches of Tanzania, 1994–1998 (quantity in tonnes, value in TSh)[8]

Year	Freshwater		Marine		Total	
	Quantity	Value	Quantity	Value	Quantity	Value
1994	228 003	30 949 458	39 072	14 001 152	267 076	44 950 610
1995	207 139	45 805 145	48 761	24 662 430	255 900	70 467 575
1996	262 572	81 209 665	59 508	38 052 517	322 080	119 262 182
1997	306 750	42 265 000	50 210	25 350 000	356 960	67 615 000
1998	300 000	47 486 100	48 000	29 273 500	348 000	76 759 600
1999	260 000	44 018 000	50 000	33 500 000	310 000	77 518 000
2000	271 000	45 500 000	49 900	32 180 000	320 900	77 680,000
2001	283 354	47 108 668	52 934	34 113 717	336 288	81 222 386
2002	273 856	54 771 300	49 674	33 372 136	323 530	88 143 436
2003	301 855	141 073 500	49 270	34 489 000	351 125	175 562 500
2004	312 040	147 743 000	50 470	40 376 000	362 510	188 119 000

Source: Fisheries Department, Dar es Salam, 2004

Table 10: Landings of herrings, sardines and anchovies in Tanzania, 1992-2002 (in tonnes)

Year	1992	1993	1994	1995	1996	1997	1998	1999	2000	2001	2002
Production	5 000	5 472	8 563	3 750	14 323	5 000	4 450	14 000	15 000	15 500	14 000

The production of the freshwater small pelagic species *Rastrineobola argentea*, locally known as *dagaa*, in Lake Victoria has increased in recent years concomitant with a steady increase in biomass. The population has increased significantly in spite of an intensified exploitation by fishermen, Nile perch and birds.

The Nile tilapia fishery is also experiencing intensified exploitation. There are no definitive data on stock biomass to indicate the current state of the stocks in the lake or to support the argument for overexploitation.

An exception to the stable marine fisheries production from 1997 onwards have been the increases in the landings of sardines, herrings and anchovies in mainland Tanzania and in Zanzibar as shown in Tables 10 and 12.

Table 11: Fish catches of Zanzibar, 2001–2004 (quantity in tonnes, value in TSh 1 000)

Fish	2001		2002		2003		2004	
	Quanti	Value	Quantit	Value	Quanti	Value	Quanti	Value
Spine foot	1 109	748 940	1 186	978 087	710	558 549	1 112	10 438
Parrot fish	1 056	553 633	828	579 016	801	601 990	758	8 006 051
Emperors	2 178	1 500	1 958	1 347	1 411	1 170	1 713	15 707
Groupers	288	173 651	268	168 803	177	208 887	618	6 390 059
Goat fish	1 739	214 074	707	252 078	363	218 281	638	5 740 757
Surgeon fish	548	132 056	684	177 013	521	178 021	589	4 754 702
Mullets	132	72 466	95	34 838	108	78 925	152	938 899
Anchovies	3 202	805 223	3 772	743 987	4 882	1 222	542	21 678
Sardine	572	184 051	899	490 124	711	241 128	1 134	9 681 659
Mackerels	1 506	900 594	1 357	966 381	1 726	1 518	7 047	8 771 574
Trevallys	511	405 865	742	581 738	472	457 300	299	10 571
Yellowfin	1 217	886 153	1 102	920 848	1 069	744 041	568	12 903
Swordfish	734	575 641	685	490 494	2 013	506 469	526	7 429 646
Kingfish	711	548 828	1 014	851 845	450	395 217	636	7 126 805
Barracuda	685	440 564	672	497 182	649	502 876	826	7 057 271
Sharks/rays	641	312 081	870	644 689	1 245	2 067	833	7 525 578
Octopus/squ	1 038	661 206	980	794 821	701	386 923	965	9 904 885
Lobsters	68	70 417	30	32 813	64	57 307	81	1 035 205
Others	2 606	1 566	2 492	1 984	2 786	2 070	3 440	31 143
Total	20 542	10 752	20 343	12 537	20 861	13 185	22 477	186 805

Source: Ministry of Agriculture, Natural Resources, Environment and Co-operatives, Department of Fisheries and Marine Resources, Zanzibar, 2004

Table 12: Small-pelagic fish catches of Zanzibar, 1998–2004 (in tonnes)

Species	199	1999	2000	2001	2002	2003	2004
Ancho	1	673	1	3 202	3 772	4 881	541
Sardine	782	1 769	971	571	898	710	1 134
Macker	959	945	1	1 506	1 357	1 725	7 046

Source: Department of Fisheries and Marine Resources, Zanzibar, 2004

SPECIES EXPLOITED BY SMALL-SCALE FISHERIES AND THEIR UTILIZATION

Marine Species

The main marine species targeted by small-scale fisheries in Tanzania are those that are large in size such as emperors, snappers, groupers, jacks, trevallys, swordfish, kingfish, sharks and rays. Kingfish, tunas, anchovies, sardines, prawns, crabs, lobster, squid and octopus are also caught. Fish landed by small-scale fishers is often of poor quality. This is due to the fact that fishermen do not carry ice onboard their vessels and fish caught far from landing sites has deteriorated by the time it is landed and sold.

There are about 14 fish landing sites along the Tanzanian coast. The Ferry Integrated Fish Market Complex of Ilala Municipality, Dar es Salaam, is the leading fish market and the landing site also receives iced fish from neighbouring districts and islands such as Bagamoyo, Mafia and Zanzibar.

The marine fish caught by artisanal fishers undergoes some form of artisanal processing depending on the target markets. Some finfishes are processed by sun-drying and salting or semi-processed by gutting and cutting in traditional ways for the local markets. Commercial species such as shrimps, lobsters and octopus are usually sold to exporting companies through the companies' agents. Shrimps and octopus are usually iced and kept temporarily in insulated containers for preservation before being processed and frozen for export.

Freshwater Species

Freshwater species harvested by small-scale fishers from lakes and reservoirs in Tanzania include Nile perch, small pelagics (*dagaa*) and Nile tilapia. The main species harvested under the name of dagaa are *Rastrineobola argentea* and *Limnothrissa miodon.* Lake Victoria and Lake Tanganyika are the main sources of freshwater *Rastrineobola argentea.* After Nile perch, *Rastrineobola argentea* is economically the most important species supporting a major artisanal fishery. Other small pelagic species in Lake Tanganyika are the two schooling *Clupeid* sardines *Limnothrissa miodon* and *Stolothrissa tanganicae.*

Rastrineobola argentea is caught exclusively by small-scale and artisanal fishermen. A labour-intensive light fishery for *Rastrineobola argentea* was started in the 1960s and expanded rapidly during the 1980s. Fishing for *R. argentea* is carried out during moonless nights. The fish is attracted by three to five lanterns. Usually, four men are operating a boat owned by someone else, who is not part of the crew. Except for the introduction of inboard and outboard engines, no further technological development has taken place. *Dagaa* is the only endemic fish species that has remained abundant in Lake Victoria since the introduction of Nile perch and Nile tilapia.

In spite of an intensified exploitation by man and predators, the *Rastrineobola argentea* stock appears to be relatively stable. As shown in Figure 4.7 below, over the last ten years, landings have been stable at between 40 000 and 48 000 tonnes per year. There is a growing demand for the product by the local and regional population as well as an increasing demand from the fishmeal industry.

R. argentea is processed primarily with artisanal methods. Direct sun-drying is the first processing step to which *dagaa* is subjected immediately after landing. Once the catch has been landed on the beach it is dried over a period of 24 to 48 hours either on the beach itself, on rocks or on grass. Women from fishing communities usually carry out this labour-intensive process. The fish has to be turned periodically to dry properly. According to artisanal fishers, the fishing ground, where the fish has been caught, apparently has an impact on product quality. *R. argentea* from rocky bottom areas has a longer shelf-life than fish caught in muddy bottom areas.

A major problem with the traditional drying process is the contamination with sand, waste water and other external agents. In addition, the absence of drying racks and shelter facilities impedes the drying process during the rainy season. Post-harvest losses, especially during the rainy season, are very high.

COMMUNITY PORTRAITS – *DAGAA* KABAJANGA BEACH

Kabajanga beach is a big landing and processing site for *dagaa* in the Mwanza region. The fishing community of 1 500 people live directly on the beach. One hundred and fifty boats with four fishermen each supply the beach with *dagaa* every day. Each boat owner usually owns more than one boat and is not directly involved in the fishing activity. The nets have to be replaced every 8 to 12 months and only one net is used per boat. The landings are not weighed or recorded due to lack of scales and fisheries inspectors at the beach.

Women dry *dagaa* directly on the sand of the beach. *Dagaa* is spread on the ground for drying and women and domestic animals walk over the drying fish. The fish processed at this beach is both for human consumption and fishmeal production. There is not enough space for drying fish, especially when large volumes are landed. Customers come from Mwanza, Kenya, Malawi and Zambia to buy directly from the artisanal fish processors. Fishmeal processors are among the biggest customers but pay lower prices. The community has not enough funds to improve the fish processing conditions, which are very unhygienic and generate high post-harvest losses. There are no storage facilities available either.

The community has expressed a need for training in better processing methods. In addition, the community would like to sell fresh fish but lack of ice and knowledge about potential demand and markets make it difficult to pursue these wishes. During the rainy season, fishermen stop their activities because the road to the major markets become impossible to travel and the community switches to agricultural activities. The rainy season lasts for six months.

Fresh *dagaa* is sold at TSh3 500 per 50 kg sack while dried *dagaa* is sold at TSh4 000 to 4 500 per 50 kg sack. The price depends on the quality of the product, which is indicated by its colour. About 2 000 sacks of *dagaa* are sold per month at Kabajanga beach.

Once the product has been dried, it is packed in 50 kg bags. *Dagaa* is dried for human consumption and for reduction to fishmeal. With regard to the processing process, there is no distinction between the processing for human consumption and the one for fishmeal.

The only industrial process, to which *R. argentea* is subjected, is fishmeal production. An increasing part of the total landings is now used for fishmeal production. Estimates vary from 50 percent to more than 60 percent. With an increasing amount of *dagaa* being caught in Lake Victoria, fishmeal and animal feed factories have obtained a new and cheaper source of supply of animal protein. Some years ago, the fishmeal and animal feed factories procured mainly low quality *dagaa,* which was not properly dried and was contaminated with sand. Increasingly, the fishmeal processing companies now demand a higher quality product. This demand generates direct competition with *dagaa* dried for human consumption.

Nile Perch

The Nile perch fishery is exclusively artisanal and divided into an inshore and offshore fishery. The inshore fishery targets small Nile perch together with tilapia, which has a ready local market. The offshore fishery targets Nile perch in deeper waters for processing factories. The fishery comprises fishermen operating traditional non-motorized fishing vessels such as canoes, using either paddle or sail power, gillnets, baited longlines as well as beach seine nets. Fishermen set gillnets and baited longlines overnight and haul them early in the morning. Those who only use longlines, set their gear during the early morning hours and haul them in the late afternoon.

After fish has been landed, it is weighed and sorted. Fish of highest quality is sold to local agents and middlemen, who sell the fish to processing factories. The factories agree on a price with the agent before the fish is bought from the fishermen. Once this has been done, the price paid to fishermen is determined by agents. In June 2005, the price of Nile perch paid by fish agents to fishermen ranged from TSh1 000 to 1 200 per kg while the price paid to fish agents by factories ranged from TSh1 200 to 1 500 per kg. According to the fishermen at Bwai beach, in most cases, agents manipulate the weighing scales to increase their profit. In some cases, agents provide equipment and capital to fishermen, who then have to sell their catches to them at lower prices. Some factories deal directly with fishermen without going through agents and provide outboard engines, nets and other fishing gear to fishermen on credit under an agreement, by which the fishermen will supply their catch to the factories.

Second grade Nile perch or factory rejects are sold to artisanal fish processors and traders. About 5 percent of the Nile perch landed is rejected by processing factories on quality grounds. Fish caught in remote places, where preservation facilities are poor, is sold fresh for local consumption or is used for artisanal processing. Other Nile perch products that undergo artisanal processing are Nile perch filleting by-products, which are bought by women from processing factories.

Drying and Salting

Sun-drying is normally used for preserving small Nile perch, which are caught illegally by fishermen or caught as bycatch when fishing for *dagaa*. Typically, the fish is dried on the beach, on rocks or on elevated racks together with *dagaa*. Just like in the case of *dagaa*, the end product is usually heavily impregnated with sand particles, unhygienic, has a rancid and bitter flavour and does not meet the expected health and nutritional requirements for human consumption.

Second grade, factory rejected low-quality fish is split, gutted, and stacked to allow salt penetration, which by itself causes a dehydration process. The process can take up to three days depending on the size of the fish. Fish is then dried under the sun individually on racks unprotected from the environment. Measurements such as quantity determination or moisture content determination are never employed. The final product is known locally as Kayabo and appears dull white to yellowish in its early shelf-life but with time develops a brown coloration with dark pigmentation. The products have a low nutritional value compared with fresh fish. Vitamins, proteins (amino acids) and lipids degrade during the course of processing and storage. The longer the product is stored, the poorer it becomes nutritionally.

Smoking

Smoking is used for factory rejects and fresh Nile perch of small size. The fish is handled more or less as in the case of salting during the initial preparations. Large size fishes are de-scaled, gutted and split while small size fishes are only de-scaled and gutted. The most commonly used kilns are Chorkor ovens. In these ovens, fish is placed on layers of racks and a fire is lit underneath. When Altona smoking ovens are used, the fish is arranged in rows on spits and a fire is lit under it. In the case of traditional Banda and Drum ovens, which usually are used by small-scale or domestic processors, only one layer of fish is placed on a rack and a fire is lit under it. Most of these traditional ovens are made of mud or mud bricks. They have limited capacity and are difficult to operate. Drum ovens while durable are also of limited capacity and difficult to operate.

With regard to the hygienic conditions of artisanal fish smoking, the environment is similar to the one described in the case of drying and salting. In the case of smoking though, the heat effect could take care of microbes if the finished product were handled hygienically. This is, unfortunately, not the case.

Other characteristics of smoked Nile perch products include the colouration of the final product depending on smoke deposition and exposure time. The final colour of the products also depends on the freshness of the raw material: the fresher the raw material the more attractive the colour. The flavour of the smoked product is a function of its freshness and the type of wood used. Texture is another important aspect. Since the temperatures used are not controlled and depend only on the combustion temperature or the source of fuel used, which in most cases is wood, the final product toughens and becomes hard to chew. Smoked products from Nile perch are mostly of poor quality because they are made from spoiled fish rejected by fish processing plants. The appearance of the product is not attractive. Smoked fish products made from Nile perch rejects are only consumed by low-income groups in the community.

Frying

Frying is used for the preservation of fresh, mostly small-size Nile perch below the minimum size limit. Artisanal fish processors buy Nile perch fat from processing factories and use it for frying the fish. Larger fishes are gutted and cut into small pieces to fit into the frying pan while fishes that are small enough to fit in the pan are fried whole. Insertions are cut across the fish to expose a maximum surface to the heat. Deep and shallow frying is carried out mostly by women only a few metres from the fish landing sites. The product is popular as it represents a ready-to-eat meal and it is usually consumed by fishermen and fish traders as they carry out their daily activities at landing sites and fish markets.

As in the case of other processing methods, hygiene problems are a major concern with this method of processing. The fact that the processing method is carried out in the open only a few metres from the landing site exposes the product to contamination from dust. Furthermore, the product is a ready-to-eat product of limited moisture content, a condition that puts it at risk of microbial and insect contamination and infection. Ideally, the product should be kept in clean protected areas and prepared in shelters that are free from dust.

Nile Perch Filleting by-products

Nile perch by-products constitute up to 60 percent of the fish after removing the fillets. Artisanal processors buy Nile perch filleting by-products such as trimmings, frames consisting of head, backbone and tail, chests, belly flaps, visceral flaps and skin from industrial processing factories. The belly flaps are dried while the visceral flaps are used for the production of oil used by local fish fryers.

The heads are split open and sun-dried and the backbone with the tail attached is in most cases fried or sun-dried for preservation. Previously, the skin was also dried for human consumption but lately the declining amount of flesh remaining on the skin due to improved filleting methods prohibits this use. The skin is now dried and used as fuel for frying. During the rainy season, drying of Nile perch filleting by-products becomes difficult because the by-products take longer to dry and a good fraction goes bad due to spoilage and insect infestation.

Considerable employment is involved in the artisanal processing of Nile perch by-products. The workers usually lack formal training and education in handling, hygiene and processing methods. Benefits from the processing and trading of Nile perch by-products trickle down to other members of the local communities. The growing fishmeal industry is a major competitor for artisanal processors of Nile perch by-products and *dagaa* threatening employment and protein supply among the poorer parts of the local population. Competition is especially strong for Nile perch frames.

Nile tilapia

Catches peaked in 2002 and then dramatically dropped in 2003. Like Nile perch, Nile tilapia is exclusively landed by artisanal fishermen. At present, there is no demand from fish processing factories for Nile tilapia. The fish is exclusively processed by artisanal processors. Nile tilapia is processed with the same methods as factory rejects of Nile perch. Most of the catch is consumed in fresh form and a part of it is smoked and fried.

CONSUMPTION PATTERNS AND MARKETING CHANNELS FOR SMALL-SCALE FISHERIES PRODUCTS

Consumption Patterns

In 1999, the world average annual per capita consumption of fish was 16 kg while the overall consumption in Africa was only 6.2 kg. For a fish producing nation, the domestic consumption of fish in Tanzania is relatively modest. In 2001, as shown in Figure 4.9, the per capita consumption of fish in Tanzania was estimated at 7.1 kg, which is only slightly higher than the all-Africa average but lower than the average per capita consumption of about 12 kg in the East Africa region, where fish has traditionally been the most affordable source of animal protein.

Various types of sardine-like fishes in dried form are an important source of protein in the traditional diet of poor and middle-income groups throughout East and Southern Africa. The availability and affordability of dried fish make it well suited for low-income households. *Dagaa* (*R. argentea*) has important implications for food security in Tanzania. While most of the Nile perch and marine fish products are exported, dried *dagaa* remains the staple fish for many households around Lake Victoria and for many communities in Tanzania.

Its high protein content and flesh composition is an advantage to consumers especially to children threatened with malnutrition. Vitamin A is particularly vital for good vision and a healthy skin. Dried fish is one of the richest sources of vitamin A. In Kenya, *dagaa* is used to prepare a protein-rich baby food based on beans, soy and maize to protect against *kwashorkor*, a disease associated with lack of protein. Access to only 10 g of dried *R. argentea* adequately addresses iron, zinc and vitamin A deficiencies common among children. In addition to human consumption, *dagaa* is also used as a source of protein for domestic animals. Fish consumption is high in the dry months of July and August when there are few vegetables that can substitute it.

Nile perch consumption in Tanzania is relatively low as most of the fish is exported. Only poor quality factory rejects and waste from the processing plants are available in local markets. Tilapia is the most consumed and preferred fresh fish by local consumers. It is easily accessible and can be found in most local restaurants and hotels, particularly those located near lakes. Most of the marine fish is consumed locally, mainly in fresh form. According to the Fisheries Department of Tanzania, almost all crustaceans and cephalopods landed are exported to the European Union, Asia and the Middle East.

Distribution and Markets

There are two major trade channels for fish in Tanzania. The first channel supplies the local market. It is commonly referred to as the artisanal or informal trade sector. The second channel supplies fish to regional markets. Fishermen sell fish to women or male traders at the landing site, who then sell the fish in nearby markets or to middlemen, who transport it to other rural markets or to distant urban markets. Traders make use of public transport, bicycles or just walk.

Fresh fish is generally traded close to the production centres and in well connected urban centres. Local street markets are still common in Tanzania even though modern retail shops and supermarkets are being established in cities. Rural areas are still penalized by the absence of proper transport, logistics and cold storage facilities.

While Nile perch and marine products such as crustaceans and cephalopods are exported to the United States, the European Union, Asia and Australia, fried and dried Nile perch by-products, dried *dagaa*, fresh and smoked tilapia and various marine products are sold in the domestic and African regional markets.

Table 13 shows that both quantity and value of Nile perch product exports from Tanzania have been increasing during the period 2001-2003.

As far as regional markets are concerned, Table 14 shows that the average fish consumption varies widely in African countries. While coastal countries like Mauritius consume an average of 22 kg of fish per person per year, the fish consumption in landlocked countries like Zimbabwe is as low as 1.4 kg per person.

Table 13: Exports of Nile perch products from Tanzania, 2001-2003 (quantity in tonnes, value in US$1000)

Product/year	2001		2002		2003	
	Quantit	Value	Quantit	Value	Quantity	Value
Belly	2 876	916	1 280	561	1 546	920
Fillets	31 386	77 212	23 829	78 233	31 561	102 375
Fish	398	68	98	15	582	98
Fish	1 452	4 689	1 011	2 857	1 354	5 774
Fishmeal	1 655	603	1 019	212	390	157
H & G	1 037	2 324	903	4 283	875	2 083
Fish	0	0	267	131	121	64
Fish skin	66	29	706	1 611	601	360
Off cuts	35	13	259	73	129	33
Fish	121	286	106	252	124	173
Kayabo	13	38	1	3	4	13
Total	39 039	86 179	29 479	88 232	37 287	112 050

Source: Fisheries Department, Dar es Salaam, 2004

Tanzanian fish is exported to other African countries in fresh or cured form. The main markets for products processed by the artisanal and small-scale fisheries sectors are neighbouring countries in Central, West and Southern Africa, the latter ones being the most important regional export market for artisanally processed fish products.

Dagaa (R. argentea) is exported to the Republic of Congo, Kenya, the Central African Republic, Sudan, Nigeria, Rwanda, Burundi, Zambia and South Africa. These markets, together with the markets of the three countries bordering Lake Victoria, i.e. Tanzania, Uganda and Kenya, have a market potential of more than 360 million consumers. the Republic of Congo is by far the single most important export destination for *dagaa*. Nile perch by-products such as dried trimmings and chests are exported to the

Republic of Congo, Burundi, Rwanda and other Central African countries. Small volumes of dried Nile perch (*Kayabo*) are exported to Senegal, Sierra Leone, Zambia and Malawi. Salted Nile perch, often in dried form, is exported to the Republic of Congo and to Rwanda.

Table 14: Per capita seafood consumption in southern Africa, 1996-2001 (in kg)

Country/ye	1996	1997	1998	1999	2000	2001
Angola	12.8	11.6	9.8	13.2	13.5	17
Botswana	3.2	4.1	3	3.5	4.7	3.6
Congo	6.5	5.9	7.1	6.5	5.8	5.9
Malawi	6.3	5.5	3.9	4.2	4.1	3.6
Mauritius	19.6	21.1	24	22.9	23.9	22
Mozambiq	2.2	2.2	2.7	2.7	2.8	2.1
Namibia	13.1	13.3	13.1	13.5	14.4	14.2
Seychelle	64.2	61.2	60.8	53.4	57.6	61.9
South	6.8	8.5	6.9	6.8	6.4	7.5
Swaziland	0.2	11.6	8.7	6.7	6	4.5
Tanzania	9.2	9.8	8.6	8	7.2	7.1
Zambia	7.6	7.2	7.5	7	7	6.5
Zimbabw	3.3	3.1	2.4	2	1.8	1.4
Average consumpti	11	12.69	12.15	11.53	11.94	12.1

South Africa represents the main export market for Tanzanian crustaceans, followed by Mauritius and Ethiopia. The main export products in 2004 were lobsters, squid and prawns. Lobsters were also exported to Zambia and Mauritius. Other important markets for Tanzanian crustaceans are countries in the Middle East and Asia.

As far as non-African export markets for fish products from Tanzania are concerned, Nile perch is the dominant fish species accounting for almost 80 percent of the value of Tanzania's total fish exports as shown in Table 15. The exported products of Nile perch include fillets, whole gutted and headless fish, fish maws and Nile perch bladders. Main export destinations for fillets are Europe, Australia, the United States of America, China, Hong

Kong SAR, Singapore, Japan and the Middle East. Sun-dried Nile perch maws are exported to Asian markets, particularly to China, Hong Kong SAR, China and Japan. Nile perch skin is exported to the UK. Other fish products exported from Tanzania to non-African markets are crustaceans, i.e. lobsters, prawns, crabs, freshwater crayfish, molluscs, octopus, squid and marine fish including small quantities of live fish. In the early 1990s, the marine shrimp production showed a promising development. The main markets for shrimp are Japan and Europe. The European market is the main importer for fish products from Tanzania.

Exports of fish and aquatic products from Zanzibar are shown in Table 16. In terms of value, lobsters, sea cucumbers and shells are the main export items.

Involvement of Women in Fisheries in Tanzania

Women occupy a central place in fisheries in Tanzania, particularly in the onshore and post-harvest sub-sector of the marine and freshwater fisheries sectors, where they play important roles as artisanal fish processors and traders. With regard to *dagaa,* the fishing and fish processing activities usually involve couples. The men go fishing during the night, bring the catch to the landing site in the morning, and women take over the processing and trading of the fish caught.

Previously, women processed Nile perch and tilapia but these species are no longer available in sufficient quantities as they are now being processed in factories. Because of this situation, many women either left the processing sector or turned to processing of Nile perch by-products such as Nile perch frames bought from fish processing factories. These women are often organized in cooperatives and most of them are single mothers, orphans and widows. In Tanzania, women handle from 70 to 87 percent of all fish trade in the artisanal sector and sell all sorts of fish products including smoked, fried, fresh and dried fish.

Many of the marketing centres, where women sell their fish, have no electricity or running water even where water pipes are installed. Despite the fact that women are of great importance to the fishing industry, they have received little attention from both government and non-governmental organizations.

Table 15: Tanzanian exports of fish and fish products, 2003

Product	Quantity (in tonnes)	Value (in US$1000)
Aquarium fish	25 pieces	249
Beche de mer	12	40
Belly flaps	1 546	920
Crabs	40	119
Dagaa	62	79
Nile Perch fillets	31 561	102 375
Fish chests	121	64
Fish frames	582	95
Fish maws	1 354	5 774
Fishmeal	390	157
Fish offal	124	173
Fish skin	601	360
H & G	875	2 083
Kayabo	4	13
Live crabs	168	926
Live lobsters	166	2 051
Lobster tails	2	12
Lobsters	146	1 443
Off cuts	129	29
Octopus	1 603	5 045
Prawns	1 672	5 980
Sea shells	896	380
Squids	298	1 240
Total	42 352	129 607

Source: Fisheries Department, Dar es Salaam

Improving Marketing Efficiency for Fish Products

There is a great potential for improving the marketing efficiency of the small-scale fisheries sector in Tanzania. Significant improvements could be achieved if better and more hygienic fish handling and storage methods and facilities were introduced. This would improve product quality, reduce post-harvest losses and increase the income of the people involved in the sector. There are also possibilities for adding value to fish products. A major problem with artisanally processed fish products is packaging and labelling. Presently, final products traded by artisanal fish traders are not packed in any form. Small pelagics are sold in cans or containers in bulk and traders use plastic shopping bags or newspapers to wrap their products. The most popular cans are engine oil cans obtained from petrol service stations. To make matters worse, the buyer generally has no knowledge of the origin of the fish and fish products he/she purchases. Improved and more eye-catching packaging and labelling could qualify dried artisanally processed products for distribution through national and regional retail chains.

Presently, about 75 percent of all traded fish go through some form of processing like freezing, filleting, smoking, canning, de-scaling and gutting before being consumed. About 40 percent of all processed fish is processed into fishmeal and fish oil and about 60 percent is destined for human consumption.

Processing of fish is one possible approach for adding value to fish, obtaining better prices, prolonging the shelf-life and giving it wider possibilities for marketing. Value-addition does not necessarily mean advanced processing but can also be achieved through improved packaging, transportation or marketing.

Enhancing product image, quality, packaging and presentation are the key factors in determining consumers' acceptance of new products and increasing marketing efficiency of products from the small-scale fisheries sector. Small pelagics in Tanzania are consumed either in fresh, in the case of marine products, or dried form as in the case of freshwater *dagaa*. The most interesting small pelagic species for value addition is *dagaa* because it represents the economically most important small pelagic species in the country. The *dagaa* industry has shown a lot of initiative in developing high quality products. One possible product is represented by smoked and spiced *dagaa*, which is currently produced at an experimental level, targeting the high income group in local markets. In Kenya, the Kenyan Fish Processors' Association is providing training for small-scale processors and traders on fish handling and sanitation and also on solar drying techniques to improve the quality and add value to *dagaa* products. Other potential opportunities for value-added *dagaa* include such products as canned *dagaa* in tomato sauce, brine and olive oil.

Table 17: Annual exports of marine products from Zanzibar, 2000-2004 (W= quantity in tonnes, V= value in TSh)

	2000		2001		2002		2003		2004	
Product	W	V	W	V	W	V	W	V	W	V
Lobster	0.0	125 750	0.3	528 000	5.5	11 001	15.0	39 764	27.	77 820
Sea cucumber	11.	13 406	25.	36 352	83.	73 486	34.3	29 640	50.	41 979
Fresh fish	1.6	7 028 300	0.5	118 500	0.0	40 000			0.0	18 000
Fish offal					0.0	5 400	3.13	1 225 000	0.1	330 000
Octopus	4.3	2 625 000	0.2	322 600	0.0	23 000	290.7	44 184	12.	13 927
Squids	0.0	78 500	0.0	14 400						
Skin of grouper									1.3	9 160 000
Shark fin	0.1	480 000	0.7	5 780 000	1.5	11 346	5.30	9 810 000		
Crabs					61.	4 260 000	2.55	1 046 000		
Shells			58.	12 660	90.	19 647	114.9	27 633	92.	32 990

Source: Department of Fisheries and Marine Resources, Zanzibar, 2004

Other small pelagic species with potential for value-addition are marine pelagics such as anchovies, horse mackerels and small mackerels. A wide range of anchovy products are produced around the world including frozen and dried anchovies, salted anchovies, canned anchovies, canned anchovy fillets and anchovies boiled in saltwater.

Common anchovy products for direct human consumption are salted anchovy fillets or whole salted anchovies, which are used, among other things, as pizza or salad toppings. More elaborate products are rolled anchovy fillets with capers or anchovy paste. Usually, salted products are packed in glass jars or easy-to-open tins while paste is offered in tubes of different sizes for catering, retail and individual use. In Malaysia and Japan, dried and spiced anchovies (*Niboshi*) are used as snacks to eat at cinemas and sport arenas.

In Tanzania, frozen small pelagics are processed in different ways. They are de-headed, split and placed in brine for one or two hours and then sun dried. Another preparation is to dip them in a sugar, chilli and pepper solution for one or two days before sun drying. The final products are used in soups, porridges, with noodles, vegetables and meat.

EXAMPLE OF VALUE ADDITION - *DAGAA* FISHERIES DEPARTMENT TRAINING CENTRE, MWANZA

The Fisheries Department Training Centre in Tanzania is experimenting with value-addition for *dagaa. Dagaa* is dried, spiced and smoked to improve shelf-life, quality and taste. The new production process takes approximately five hours while the traditional, artisanal processing takes 24 to 48 hours. The product is of a high quality and sold to local communities, schools and institutions. There is a great demand for this high-quality product.

There is need for investment to expand the production capacity. Currently, only about 100 kg are produced per day. A strong potential for exports to other African countries is foreseen. The local population would benefit through higher income and improved product quality. Post-harvest losses could be reduced significantly.

Potential Value-added Products

Various studies have been conducted to assess the possibilities of producing value-added products from Nile perch by-products through the preparation of fish burgers and sausages for export markets and by using the waste products for the production of pet food. While the proposed products would improve the yield and increase foreign exchange earnings, such innovative value-added products from Nile perch by-products for exports could be detrimental to the existing artisanal industries and markets in East Africa. The existing market for fried Nile perch frames and the regional market for Nile perch trimmings, chests and oil are well established and support food security for a number of low income and vulnerable people in Southern and East Africa.

Introducing higher quality and hygiene standards for the present product forms would be the first step in the right direction. Current processing methods suffer from a lack of hygiene and basic equipment. Simple facilities like drying racks or more efficient cooking facilities could reduce post-harvest losses. Training in improved fish handling and hygiene would have a positive impact on the quality and nutritious value of the final product. In addition, improved and more eye-catching packaging and labelling could increase customer acceptance and distribution in national and regional market chains. Vacuum packaging of fresh, fried and smoked products such as whole Nile perch and Nile tilapia would contribute to a longer shelf-life.

Target Groups for Value-added Products

Value-added products from small-scale producers could target two different types of consumers. The first group is represented by the current target group, i.e. low income consumers in domestic and regional markets.

The products to be sold to this group are high-quality products but without extra ingredients added, e.g. spices and salt, that would increase the price as long as the product is produced in hygienic conditions and fit for human consumption.

The second potential target group is represented by the economically well-off. In Africa as a whole, there is a growing middle class, mainly living in urban areas, which is able to pay for quality products. This type of consumer is looking for attractive packaging like pull-top oval tins or transparent plastic containers with foil tops.

Intraregional trade in the SADC region is currently underdeveloped. In 2001, an estimated 150 000 tonnes of fish products were distributed within the region. This represents only 10 present of the 1.5 million tonnes of fish produced annually. The trend is likely to change as better trade agreements are implemented. The SADC Protocol on Fisheries is of particular importance to the cross-border fish trade by small-scale traders because it promotes intraregional fish trade through the reduction of trade barriers and increased investment within the fisheries sector.

Also, the recent trends to expand international trade through tariff reduction and trade agreements promoted by the World Trade Organization is of particular importance for Tanzanian marine products such as shrimp, prawn and lobster, which are exported to international markets. Global international trade in fish and fish products is growing. A recent study shows an increase in international fish trade from US$52.7 billion in 1996 to a record US$63.3 billion in 2003.

Constraints and needs

Constraints

Small-scale and artisanal fisheries are confronted with serious problems throughout all stages of fishing, fish processing, distribution and marketing. The major obstacles are listed below.

Security

The majority of fishermen fishing at sea and in major lakes have expressed concerns about theft and lack of security while fishing. Fishermen are being attacked and even killed by thieves trying to steal their gear and boats. The most dangerous place appears to be Lake Victoria, where security is rapidly deteriorating.

Lack of Credit

Artisanal and small-scale fishers in remote areas of Tanzania experience difficulties in obtaining credit to purchase fishing equipment. Currently, fishermen rely on factory owners and agents to provide them with fishing equipment in exchange for supplying fish at lower prices. Women in particular are affected by the lack of credit in their communities. Providing credit to women would enable them to buy processing facilities, increase production and access new markets both locally and regionally. Despite the fact that women have a higher loan repayment rate than men, they still do not have access to credit.

Lack of knowledge of and infrastructure for handling, storage facilities, transport, and distribution There is a lack of sufficient knowledge of and facilities for fish handling and preservation. Currently, post-harvest losses due to poor handling are estimated to be as high as 40 percent of total fish landings. Products processed at the artisanal level are usually prepared in unhygienic conditions with negative impacts on the quality of the final product. The availability of simple cold storage facilities and the application of hygiene standards in processing areas for fish products could easily reduce this predicament and subsequently alleviate the pressure on fish stocks through a more efficient utilization of the available resources.

Poor transport system is hampering the distribution and marketing of fish in Tanzania. Most of the road networks in rural areas are in poor condition and cannot be used during the rainy season. Flights between African countries are very expensive. A functioning transport system including proper roads and the use of insulated trucks could facilitate the distribution of fresh fish in domestic markets and to neighbouring countries.

Taxes and Import Duties

High tariffs, a lack of harmony of currencies and inefficient customs services have prevented most African countries from increasing trade between them. The volume of cross-border fish trade in Tanzania is very difficult to assess as most of the trading is done informally and in small quantities to avoid taxation. Many efforts are being made by a number of regional and sub-regional organizations to identify a common trade policy for their respective member countries but so far very little has been achieved and countries have not enforced regional trade initiatives. Some African countries are also afraid that lowering or eliminating tariffs on trade with their regional partners will deprive them of an important source of government revenue.

Lack of Information and Education

Many small-scale fishing communities are located in small isolated settlements along the shores of coastal and inland waters and have no access to information. Illiteracy is widespread in fishing communities. This hampers their access to appropriate technology for fish preservation and processing as well as their ability to access information on resources, markets and prices. The region is also characterized by poor dissemination of information on appropriate techniques of fish handling, preservation, processing and distribution methods.

Needs

Improvement of product quality and marketing efficiency for small-scale fisheries products will require financial investments, which governments and local communities are unable to make on their own. The fisheries sector requires external finance and technical assistance from donor agencies in areas such as training, research and infrastructure. Training and technologies for fish handling, preservation, processing and storage as well as quality assurance systems are needed to add value to the current products and increase commercialization of the small-scale fisheries production to benefit fisherfolk through higher prices for better quality fish products.

Infrastructure for fish processing and marketing and construction of laboratories for quality assurance are urgently needed. Fish processing shelters would be beneficial especially during the rainy season when processing in major landing sites closes down. Processing facilities such as improved drying racks for drying *dagaa* and Nile perch are needed to improve hygiene standards.

Fishing communities need financial assistance to establish solar drying facilities as currently used in Kenya. Solar drying techniques will improve the quality of the end product, increase production and relieve women of the current labour-intensive methods of fish processing. In addition, solar panels and more efficient cooking facilities are required for frying Nile perch and Nile tilapia to eliminate the use of firewood, the collection of which has increasingly contributed to environmental degradation in most of the rural communities.

Clean tap water is missing in many processing and trading areas. Existing marketing infrastructure needs to be rehabilitated and new ones need to be built in many of the fish markets following the examples of the new Kirumba market in Mwanza and the Ferry Integrated Fish Marketing Complex in Dar es Salaam both financed by the Japanese government. The Kirumba market consists of a jetty for landing of fish, two shelters and wooden racks for wholesale and retail trading. This marketing structure was specifically established to cater to small-scale traders supplying regional and domestic markets. The new fish market in Dar es Salaam, which is the biggest fish market in East Africa, has seven sections including a fish landing site, an auction building, fresh fish retailer stalls and a processing section for frying fish.

Road rehabilitation and maintenance is another major concern. Most of the roads connecting fishing communities are in bad condition and need to be rehabilitated in order to increase access to markets both in terms of selling products and buying commodities. In addition, post-harvest losses have been encountered due to the difficulties of transporting fresh products destined for inland markets because of poor roads.

Interest has been shown by small-scale fishers to have ice-making machines, insulated trucks and storage facilities available at landing sites. The ice could be used onboard the vessels to preserve catches and extend the shelf-life of fresh fish, while insulated trucks would be used for distribution of fresh fish to markets.

Fishermen need credit facilities where they can borrow money to purchase fishing equipment. Most of the landing sites and processing plants have community-based organizations such as Beach Management Units (BMU) and cooperative societies with chairpersons, treasurers and secretaries. Credit facilities should be provided to and administered through these community-based organizations. There is also a need for the promotion of village level savings associations accompanied by training on savings schemes and creation of awareness of the importance of savings for income generation.

Markets for Fish and Fish Products from Small-scale Fisheries in Latin America

ARTISANAL FISHERIES AND AQUACULTURE IN LATIN AMERICA

In Latin America, the term "artisanal fisheries" is commonly used in many countries to distinguish this sector from industrial fisheries. Different countries have different definitions. In Brazil, it is not the size or dimensions of the fishing gears used but the working relationship, which defines artisanal fisheries. An artisanal fisherman is self-employed and owns his fishing gears. He works together with relatives or associates without a formal employer-employee relationship.

In Mexico, the terms "small-scale" and "artisanal fisheries" are not commonly used. When used, they distinguish fishing and production areas. Small-scale and artisanal fisheries operate within three nautical miles from the coastline as opposed to high sea fisheries and use small boats defined as boats of less than 10 GT. Boats with more than 10 GT are categorized according to the species targeted, i.e. tuna boats, sardine and anchovy vessels, shrimp vessels and white fish or multiple-purpose fishing vessels.

In Peru, artisanal fisheries are defined as those practiced by boats having a fish hold capacity of up to 32.6 m^3, which corresponds to a vessel size of about 30 tonnes and a vessel length of up to 15 m. Artisanal fisheries are also defined as fisheries, which catch fish for human consumption and not for reduction to fishmeal. There are different landing sites for artisanal fishing boats and industrial fishing vessels even in the same harbour.

All other Latin American countries have their own definitions of small-scale and artisanal fisheries related to the size of fishing gears, fishing areas and employment relationships and conditions. In some countries, the differentiation between artisanal and industrial fisheries is more important than in others according to the importance given to this differentiation in national fisheries statistics and other factors.

In the light of current global concerns, the characterization as artisanal fisheries should be seen as positive characterization. The limited coastal fishing grounds accessible to and exploited by artisanal fisheries for instance highlight the important potential role artisanal fishermen and their communities can play in coastal management. The limited size and dimensions of their gears have a positive impact on the sustainability of fishery resources exploited by them. The free association of skilled workers of a community without formal employment relationships can be seen as a very modern way of working and of creating wealth.

The semantic differentiation between "small-scale" and "artisanal" may have implication for the perception of the public in general as well as of consumers. In some countries, for food products including fish, the seal "artisanal product" is perceived as a seal of quality. In Latin American countries, with very few exceptions such as the Brazilian whale hunting in the eighteenth century, all fisheries could be considered as being artisanal until the middle of the twentieth century. Artisanal fisheries are therefore largely also traditional fisheries.

By the nature of their activity, for joining efforts and for common security, fishermen in almost all countries live in separate communities. Fishermen villages along the coast or along rivers or fishermen districts in big coastal towns and cities are normally established around small wharfs or landing facilities. Since the first half of the twentieth century, most of these communities are organized in associations or cooperatives. The organization of fishermen communities in the first half of the twentieth century was supported by the national navies as a means of supervising national coastlines through "the eyes and ears" of the fishermen.

It was in this sense that the Brazilian Navy, for instance, organized a four-year (1919-1923) mission of a cruiser to organize "fishermen colonies" with basic medical care and schools and a strong sense of patriotism. Most of these colonies still exist today.

The modern Latin American industrial fishery sector was developed during the 1960s and 1970s targeting mainly export markets. The development of industrial fisheries focused on shrimps, lobsters, tuna, hake and other commodities demanded by growing international markets. Other industrial producers, however, focused more on domestic markets such as the suppliers of canneries and some processors of frozen seafood.

Artisanal fisheries have improved during this period through the modernization of boats and gears, utilization of bigger boats, a growing use of steel or fibreglass in substitution of wood and the progressive substitution of oars and sails through inboard and outboard motors. Artisanal fisheries, however, remained largely focused on the domestic markets, which were growing everywhere, following a rapid population growth in most Latin American countries combined with growing urbanization. About 80 percent of the Latin American population is currently urban and concentrated in big cities.

Together with the development of fisheries as part of the overall development of the countries, road networks were constructed along most of the coasts, facilitating the transport of catches to markets and also encouraging tourism, which began to occupy large portions of the coastlines through the construction of beach residences for local urban middle and upper classes and hotels for domestic and foreign tourists.

The impact of this development on traditional fishing communities was multiple. On the one hand, it made the sale of their fish production easier through better transport facilities and by bringing consumers closer to fishing communities, at least during vacation times. On the other hand, many fishermen were attracted to other activities like construction industries and tourism. Many fishing communities, particularly small ones, vanished and were absorbed by new employment opportunities. Other fishing communities quickly adapted to the new times and modernized their fishing activities.

The cases of Mexico, Peru and Brazil illustrate the general trend of the Latin American continent. These three countries concentrate more than half of the Latin American population and also more than half of the Latin American fish production. The largest part of this production is of Peruvian *anchoveta* processed into fishmeal as shown in Table 1.

The Case of Mexico

There are about 1 650 fishermen cooperatives in Mexico known as *Sociedad Cooperativa de Producción Pesquera.* In the 1940s, the Federal Government had promoted these cooperatives and given them for many years the monopoly of the catch and marketing of eight important species including shrimps, lobsters and oysters.

The case of Tamiahua Fishermen's Cooperative is illustrative of how artisanal fishermen can organize their community. The town of Tamiahua is located at the southern mouth of the Tamiahua lagoon. The lagoon covers an area 88 000 ha and is about 100 km long and 25 km wide.

The lagoon serves both tourism and fisheries. Catches in the lagoon are regulated through concessions given to local fishermen organizations. These concessions ensure the protection of habitats and the conservation of the aquatic and natural resources of the lagoon. The fishermen organizations have agreed to use only selective fishing gears, which ensure the sustainability of their fishing activities and restrict possible negative ecological impacts. The main species harvested in the lagoon are oysters, shrimps and some white fishes.

The town of Tamiahua has 5 153 inhabitants, which account for 19.2 percent of the population of the entire municipality. The town has 2 562 registered fishermen, of whom 343 belong to the Cooperative "Fishermen of Tamiahua". This cooperative was legally established as a cooperative society with limited responsibility and variable capital. The cooperative is one of the biggest social organizations in the state of Vera Cruz. Since its establishment in 1972, the number of members of the cooperative has remained the same. The cooperative has concessions for the extraction, catch and processing of oysters and shrimps, as well as authorizations to catch blue crabs and finfishes such as seabreams, snooks, mullets, sea trout and other species inside as well as outside the lagoon. The concessions are renewable every 20 years.

Table 1: Average seafood consumption in Latin America and the Caribbean, 1999-2001

Country	Production (tonnes)	Non-food use (tonnes)	Import-export balance (tonnes)	Population (1 000)	Per capita production (kg)	Per capita consumption (kg)
Mexico	1 366 071	238 725	- 92.070	98 928	13.8	10.4
Peru	9 032 601	8 302 807	-212.085	25 950	348.1	20.1
Brazil	922 658	60 635	229.241	171 795	5.4	6.5
Total Latin America & Caribbean	19 197 660	12 617 204	-2.147.991	520 214	36.9	8.7
Share of 3 countries of total Latin American production, nonfood use and population	59%	68.2%		57%		

Source: FAO, 2002, Fisheries Statistics – commodities

Production Techniques in Tamiahua

Oysters are harvested with fibreglass boats of 18 to 23 ft (5.5 to 7 m) in length and outboard motors of 15 to 40 hp. Grapples are used, which consist of two rakes with 26 nails joined in the middle and operated like pincers. On banks with high oyster density, it is possible to extract around 200 oysters in one operation. Oysters under 8 cm in length are returned to the oyster bank. The oysters of a proper size are packed in bags of 35 to 40 kg. These bags may contain up to 500 oysters each.

Shrimp is caught from 19.30 hours in the evening to 05.30 hours the following morning by only one person per boat operating a scoop net in a wooden corral locally called "charranga". To catch blue crabs, a crew of two persons in a boat operate crab pots. Pots are set with fish or chicken parts as bait for about 30 minutes.

Finfish are caught with gillnets of about 300 m length, 3 m depth and mesh sizes between 76 to 102 mm. The nets are made of mono- or multifilament twine of 0.2 to 0.55 mm. Nets are set for 8 to 12 hours. A boat fishing for finfish has a crew of 2 or 3 persons. The main species caught are seabreams, snooks, mullets and sea trout. The catches are kept onboard in fish boxes with a capacity of 40 to 50 kg without ice. The boats return in the evening to unload their catches at the landing site, where fishes are kept in basins covered with wooden panels.

Fishermen bring their catch to the cooperative to be sold locally or to be sent to other towns of the region and to Mexico City. In the latter cases, the catch is placed in insulated containers made of fibreglass and kept in ice. The ice is bought at the ice plant of the town at a cost of 90 to 100 Mexican pesos, which is equivalent to US$8.40 to 9.50 per ice bar. The monthly average cost of ice incurred by the cooperative is about US$2 000. Table 5.2 and Figure 5.1 show the catch record of the Tamiahua Fishermen's Cooperative society over from 1996 to 2005. This period includes three years, i.e. 1999 to 2001, when crab catches were banned. While the catch records of shrimp suggest that the resource is exploited in a sustainable manner, the oyster banks seem to have been overexploited during the period from 2000 to 2004. It seems that the exploitation of oysters in the lagoon is not well managed.

Table 2: Fish landings in Tamiahua, Mexico, 1996-2005 (in tonnes)

Species	1996	1997	1998	1999	2000	2001	2002	2003	2004	2005
Oysters	357.28	4.48	22.99	27.47	212.94	493.43	1 102.11	365.64	199.28	12.07
Shrimps	148.60	114.24	127.92	182.18	132.820	170.07	185.17	126.06	250.16	177.07
Crabs	13.64	.90	1.97	0	0	0	17.34	12.58	36.46	26.81
Finfish	255.97	176.79	124.53	44.81	41.70	9.39	50.44	49.07	97.61	44.74

Source: Tamiahua Fishermen's Cooperative

As shown in Table 3, the cooperative sorts, processes, packs and sells the production of its members. The preparation of the final products, including packaging and the peeling of shrimp, is done by the women of the community. Women do much more than simply process fish. They are perceived by the entire community as the real managers of the earnings of their fishermen husbands and manage their homes and families including the economic affairs.

In November 2005, the oysters produced by the Tamiahua Fishermen's Cooperative were sold in Mexico City and Guadalajara, while shrimps were mainly sold in Mexico City and Tampico. Crabs and finfish were sold locally.

The Cooperative sells most of its products to local wholesalers because it still lacks the means to sell the products directly to retailers and consumers. The local wholesalers have better possibilities than the cooperative to store and transport large quantities of fish products to markets. The cooperative has only one truck, which is not sufficient to transport the entire production. Similarly, the cooperative does not have enough space to display products for local retailers. It works therefore with the 20 independent local fish stores of Tamiahua, which take charge of selling its products locally. There is no supermarket in Tamiahua. The closest supermarket is in Tuxpan, 40 km from Tamiahua. It belongs to the Chedrahui chain and buys weekly 100 kg of big shrimps, 50 kg of small shrimps and 100 kg of finfish directly from the cooperative.

Table 3: Products sold by the Tamiahua Fishermen's Cooperative, Mexico

Product	Description/presentation	Packagin	Perceived quality
Fresh fish,	Variety of species, washed and packed in plastic boxes with	None	Fresh, good flavour, nutritious, affordable
Blue crab,	Whole, fresh	None	Fresh, good flavour, nutritious, affordable
Crab meat	Cooked	1 kg nylon	Good flavour, nutritious,
Shrimp	Blue and "coffee" shrimp, packed in plastic boxes with crushed ice	None	Fresh, good flavour, nutritious, affordable
Oysters	Shell-on	45 kg bags	Fresh, good flavour, nutritious, affordable
Oysters	Shell-off	Bags	Iced, good flavour, nutritious, affordable

Retail prices for finfish and crabs in Tamiahua are normally equal or sometimes higher than the prices at the Nueva Viga wholesale market in Mexico City. Higher prices for shrimp and oysters in Mexico City, up to 35 percent, make it logical to sell these products there. Compared with the prices at the wholesale level, fishermen in Tamiahua receive fair prices for their production as can be seen in Table 4. The relatively high prices paid to fishermen are one of the positive results of joining efforts in a cooperative.

In addition to the fair prices paid to fishermen, the cooperative also earns revenue from its sales. Based on the prices and gross margins shown in Table 4 by species and the volume of catch during the first nine months of 2005, the gross earnings of the cooperative during the first 9 months of 2005 can be estimated about US$200 000. During its 33 years of existence, the Tamiahua Fishermen's Cooperative has managed to survive, to benefit its members and indirectly the whole community of Tamiahua. This is due to the traditional support given by the Mexican government to social organizations and also to the seriousness of management of the cooperative during all those years.

The monitoring of the present situation, the identification of problems and opportunities and the development of strategic plans shown in Table 5 make the cooperative look like any professionally managed medium-sized enterprise in the fishery sector.

Table 4: Fish prices and profit margins at Tamiahua, Mexico, 2005 (in Mexican pesos)

Species	Price paid by cooperativ e to	Price paid by local wholesaler or retailer to	Margin of coopera tive	Price paid by local consumer	Margin of wholesalers/ retailers (c/b, in %)
Lisa	16.00	19.00	18.7	21.00	10.5
Sargo	35.00	37.00	5.7	40.00	8.1
Lebrancha	7.00	10.00	42.9	11.00	10
Robalo	70.00	75.00	7.1	85.00	13.3
Chucumite	30.00	33.00	10	35.00	6.1
Trucha	30.00	33.00	10	34.00	3
Mojarra	30.00	33.00	10	35.00	6.1
Trucha	15.00	20.00	33.3	25.00	25
Tilapia	10.00	15.00	50	19.00	26.7
Ostión*	90.00	100.00	11.1	125.00	25
Cam arón	85.00	95.00	11.8	120.00	26.3
Camarón	55.00	65.00	18.2	85.00	30.8
Jaiba cruda	20.00	22.00	10	25.00	13.6
Palota	32.00	36.00	12.5	38.00	5.6
Cazón	15.00	20.00	33.3	26.00	30
Bacalao	24.00	26.00	8.3	30.00	15.4
Bandera	10.00	11.00	10	13.00	18.2
Cubera	27.00	27.50	1.9	30.00	9.1
Chabela	27.00	28.00	3.7	30.00	7.1
Chopa	10.00	12.00	20	14.00	16.7
Churro	3.00	4.50	50	6.00	33.3
Guachinang	55.00	63.00	14.5	75.00	19
Gurrubata	8.00	11.00	37.5	14.00	27.3
Jurel	12.00	15.00	25	19.00	26.7
Pargo	25.00	30.00	20	38.00	26.7
Pámpano	30.00	35.00	16.7	40.00	14.3
Raya	8.00	11.00	37.5	15.00	36.4
Rastrero	6.00	8.00	33.3	10.00	25
Ronco	5.00	7.00	40	10.00	42.9
Sierra	11.00	13.00	18.2	15.00	15.4
Tonton	8.00	11.00	37.5	14.00	27.3%

Source: Tamiahua Fishermen's Cooperative, US$1 = Mexican pesos 10.66 (in October 2005)

The main difference between cooperatives and private companies, however, is the fact that the cooperative is owned by its members, who effectively work and produce a common wealth from a common resource.

Table 5: Strategies and goals of the Tamiahua Fishermen's Cooperative

Strategy	Impact	
Harvesting of fish and shellfish		
Modernizing equipments	Improved quality	■ Develop research activities related to inland waters through consulting firms and research centres ■ Obtain financing options for fishermen ■ Introduce support
Processing		
Develop processin	Add value to products	■ Search for financing options for
Marketing		
Better positioning of products in the regional market	Increase the demand for products, increase selling prices, improve the image of products	■ Implement market surveys through consulting firms ■ Implement a publicity campaign in order to better inform consumers ■ Establish different
Transport		
Acquire more adequate transport	Reduce losses, increase opportunities of delivery, improve quality of products	■ Identification of better conservation and transport equipment

The Tamiahua Fishermen's Cooperative can be considered as a successful artisanal fishermen initiative in Latin America.

It can also serve as an example for many others even though there is still much that can be improved.

The Case of Peru

In Peru, 62 341 persons are involved in artisanal fisheries and fish farming activities. As shown in Table 6, the vast majority of these, i.e. 87 percent, are directly involved in marine and freshwater capture fisheries, while 12 percent are involved in fish farming. Only one percent of all persons employed in the artisanal fisheries and fish farming sector are involved in fish processing.

Artisanal fishermen in Peru are organized in some 300 fishermen associations. They land their catches at 109 marine and 13 inland fish landing sites called "caletas", which are exclusively reserved for artisanal fishermen. The artisanal fishing fleet of Peru consists of 6 258 boats, which fish up to 80 nautical miles offshore. In 1999, 237 881 tonnes of fish were landed by the fleet, most of which were destined for the domestic market and for direct human consumption.

Table 6: Geographical distribution of artisanal fishermen in Peru

Mari ne	Number of artisanal fishermen	Number of artisanal fishing boats
Tumbes (Pto Pizarro -	2 125	468
Piura (Máncora –	3 576	911
Piura (Colán –	3 631	834
Piura (Parachique –	1 896	455
Lambayeque (San José	2 938	285
La Libertad (Pacasmayo –	1 080	172
Ancash y Norte Lima	3 299	784
Lima (Calta Vidal –	2 146	741
Lima Centro y Sur (Chucuito – Cerro	1 440	474
Ica (Tambo de Mora –	2 372	626
Arequipa	2 318	260
Ilo - Tacna	1 177	248
Total	27 998	6 258

Source: MIPE (Vice-Ministry of Fisheries for Peru), 2000

Artisanal fishermen and fish landing sites are located along the entire Peruvian coast. Table 5.6 presents the geographical distribution of artisanal fishermen in Peru. The density of artisanal fishermen and fishing boats is slightly higher in the northern than in the southern part of the country. The average crew size of an artisanal fishing boat in Peru is between four and five fishermen.

The National Directorate of Artisanal Fisheries under the Vice-Ministry of Fisheries encourages artisanal fishermen to join or form social organizations, i.e. associations, unions or companies. There are currently 108 of such organizations functioning in Peru.

The fishing community of El Chaco, located in the district of Paracas, 18 km south of the city of Pisco and 260 km south of Lima, is a good example of the variety and complexity of Peruvian artisanal fisheries. In this desert area crossed by the southern Pan American highway, the first settlers appeared in the 1960s and started fishing activities as a means of subsistence. Since then, the population of El Chaco has not grown much and even started to decline after 1980. National statistics record a population of 727 persons in 1961, 1 378 persons in 1981 and 1 196 in 2000.

The fisherfolk community of El Chaco is located close to the town of Paracas in an important tourism area, which includes the National Park of Paracas. The park was created in 1975 as a reserve for biodiversity and ecosystem reserve. The park is also an archaeological site and has pre-Incan relics. The park is the only marine protected area in Peru. Research in the park is carried out by national and international non-governmental organizations (NGOs) like Pronaturaleza, ACOREMA, the Nature Conservancy and WWF. These NGOs have implemented training programmes for fishermen on resource management. The Humboldt Current ensures a rich primary marine life. The coast has one of the largest Peruvian natural scallop banks known as "La Pampa".

The community of El Chaco has 190 registered fishermen, of whom 165 are directly involved in fishing. The other 25 fishermen of El Chaco do not go out fishing but work at the local wharf instead. As shown in Table 7, the fishermen operate 46 boats. More than two-thirds of the boats are powered by outboard engines while the rest is powered by inboard engines. About half of the boats have a fish hold capacity of between two to three tonnes while the other half of the fleet has a fish hold capacity of less than two tonnes. The fishermen operate traditional fishing gears such as gillnets, purse seine nets and longlines.

Table 7: Fishing fleet of El Chaco, Peru

Fish hold	Number of	Type of engine	Number of
0.5 to 2 MT	24	Outboard	32
2 to 5 MT	22	Inboard	14
Total	46	Total	46

Source: Vice-ministry of Fisheries of Peru

Many of the fishermen are also divers, who dive for scallops. The productivity of the natural scallop banks is due to El Niño. In years when El Niño does not appear, the scallops grow in deeper waters and can no longer be accessed by divers. This has encouraged the development of aquaculture based on natural seed collection. Scallops are grown on lines suspended from floating frames. The fishing community of El Chaco also harvests algae, mainly edible *Rhodophytes*. The landings of fish, shellfish and algae at El Chaco from January to August 2005 are shown in Table 8. In terms of quantity, anchovy is the most important species landed followed by scallops and algae.

Table 8: Fish and shellfish landings at El Chaco, Peru, January-August 2005

Species	Volume (in tonnes)
Scallops	655.6
Top shell	132.5
Abalone and mussels	18.4
Algae	531.8
Anchovy	5 215.8
Silverside	142.7
Mackerel	64.0
Horse mackerel	46.2
Limpet	1.1
Octopus	25.0
Mahi-mahi	20.0
Total	6 853.2

Source: Vice-ministry of Fisheries of Peru

Anchovies and other pelagic species are sold to processing plants in the Pisco region, where they are salted and marinated. A boat owner has typically two possibilities to sell his catch. He can either sell his anchovy catches to a processing plant and other species to general fish wholesalers or he can offer the processing plant to lease his boat and its crew for a fishing trip. In this case, which is less common, the plant covers all operating costs, rewards the crew and keeps the entire catch.

Other fish species are bought by wholesalers at the wharf and transported to Pisco and Lima, where they are sold to other wholesalers or directly to retailers. Sometimes, supermarket chains from Lima send refrigerated trucks and buy fish directly at the wharf. However, unlike traditional wholesalers, these buyers normally do not pay in cash. Table 9 shows the prices of selected fish and shellfish species as well as the profit margins of wholesalers, local markets and supermarkets.

The profit margins of wholesalers and supermarkets buying from the fishing community of El Chaco, Peru, are much higher than the profit margins of wholesalers and retailers buying from the Fisheries Cooperative of Tamiahua, Mexico, as shown in Table 4. In the Peruvian case, it can be assumed that fishermen could get much higher prices for their catches if they were organized in a cooperative, used ice for the preservation of their catches and were able to attract more wholesalers to increase competition.

Women of fisherfolk families are involved in local retailing. They usually gut and clean fish and chuck shellfish, which they sell locally and also in Paracas. They also prepare and sell local fish-based dishes like the traditional "ceviche" or "chupe", a local soup. Women contribute significantly to the income of fisherfolk families. Women are also employed as workers in the processing plants of the region. Their work normally consists of gutting, de-heading and cleaning fish, mainly anchovies, which are marinated. During the tourism season in summer, women and children make handicraft items from shells or sea lions' teeth. Women of fisherfolk families also offer their homes as accommodation for tourists.

The community of El Chaco still lacks a fishermen association or cooperative. The scallop divers have joined the Shellfish Harvesters Association from Pisco, which has 65 members from both Pisco and El Chaco. As far as the future of artisanal fisheries in El Chaco is concerned, the small community of less than 200 active fishermen is likely to decline because of the further development of tourism and related employment opportunities. It is doubtful that the artisanal fisherfolk community will be able to survive for long without organizing itself and without stronger government support, be it from the municipality, from the province or from the Directorate of Artisanal Fisheries of the Vice-ministry of Fisheries.

Table 9: Prices and profit margins of selected fish and shellfish species landed in El Chaco, Peru, 2005

Prices in nuevos soles/kg[1]

Species	Wharf (a)	Ventamilla wholesale market in Lima (b)	Margin b/a (in	Local markets (c)	Margin c/a (in	Supermarkets (d)	Margin d/a (in
Silversid	0.50	1.70-2.20	290	3.0-3.50	550	4.00	700
Drum	0.50-	1.50-2.00	218	2.50-3.0	400	3.50	536
Snapper	0.50-	1.50-2.00	218	2.50-		3.50	
Horse	2.00- 2.20	3.50-4.00	79	4.50	114	5.00	138
Blenny	2.50-	5.00	82	6.00-	127	7.00-7.50	164
Crab[2]	4.00-	8.00	78	10.00	122	12.00	167
Scallops[3]	10.00	18.00	80	22.00	120	35.00	250
Mussels	3.00-	8.00	100	10.00	150	12.00-	237
Top shell	3.50-	8.00-9.00	100	10.00-	147	12.00-	218

Source: direct observation at Maria Ayala, November 2005
[1]US$1 = S/.3.40

[3]per "manojo" (= 8 dozens)

El Chaco is an example of a great number of artisanal fishing communities in Peru, which have so far been unable to take advantage of increasing tourism, by expanding their fisheries activities through the opening of seafood shops, restaurants and similar activities. The main opportunities for the fisherfolk community of El Chaco are based on the fact that the area is classified as a national park. Environmentally friendly activities such as scallop aquaculture, improving the quality of fish landed by using ice onboard their vessels and better marketing of their catches could result in a revitalization of the local community and economy.

The Case of Brazil

At the end of the First World War, the Brazilian Navy helped to establish fishermen colonies. There are currently 750 fishermen colonies in Brazil, organized into 23 federations, one for each of the 24 federal states of the country with the exception of one state. The total membership of the fishermen colonies and their federations stands at 326 696 fishermen.

While most of the fishermen colonies are associations, 50 of them are organized as fishermen's and fish farmers' cooperatives. By the end of 1999, the Brazilian Cooperatives Organization (OCB) consisted of 14 registered cooperatives working in the fields of fisheries and aquaculture. The National Association of Fisheries Cooperatives (ANACOOP) counted 53 associated fisheries cooperatives and the Ministry of Agriculture counted 90 formally registered cooperatives, many of which, however, were in a precarious state or not functioning at all. As far as fish farming is concerned, the 278 128 tonnes of farmed seafood produced in 2003 were produced by 19 277 registered fish farmers.

The case of the Women's Association of Betume in the northeastern state of Sergipe is an example for a type of association, the scope and purpose of which extends beyond production. The village of Betume where the association is located has 900 inhabitants and is located in the municipality of Neópolis. The villagers traditionally earn their living from rice culture and cattle ranching. As Betume is located at the lower São Francisco River, the region was chosen at the beginning of the 1990s by the Federal Government of Brazil for the development of tilapia aquaculture and other freshwater fish culture, mainly the culture of *Colassomas*. Many rice growers of the lower São Francisco region became small-scale fish farmers. Each fish farmer produces from 20 to 30 tonnes of fish per year. The fish farmers formed associations and cooperatives, which, unfortunately, had neither ways and means of processing their fish nor of selling it in markets other than local markets.

In 1997, a group of 14 women, wives and daughters of local farmers, established the Women's Association of Betume with the purpose of adding value to farmed fish by filleting it. With the help of the federal organization CODEVASF, the women built a small fish processing establishment equipped with stainless steel working tables, working tools, water and electricity. With the help of an international cooperation project, funded by the Common Fund for Commodities and implemented by INFOPESCA, the women installed a small ice plant with a production of three tonnes per day. The ice was used to preserve raw material and processed products and was also sold to local fishermen, fish farmers and restaurants. Through the sale of ice, the association increased its working capital and could procure larger volumes of fish for processing.

The farm gate price of tilapia in the region is from R$2 to R$2.4 per kilo. The association sells tilapia fillets at R$ 11, in most cases directly to restaurants in the region as well as in Aracaju, the capital city of the state of Sergipe. Considering the fact that 3kg of whole fish are required to produce 1kg of fillets, the gross margin of the association per kg of fillet ranges from R$3.8 to R$4.4. The break-even point is reached at one tonne of fillet per month after paying the national minimum salary of R$300 to each of the 14 members of the association.

The selling price of the association is relatively high because of the high quality of the product. The quality is assured from the moment the fish is taken out of the water by preserving the fish in ice. The final products, the fresh fish fillets, are wrapped in plastic and kept in ice. Associated with the quality of the product are the logistics of supplying clients, who are willing to pay for quality, rapidly and regularly. In this sense, the choice of the association to supply restaurants rather than supermarkets or municipal markets was very adequate. The association also encourages villagers to farm tilapia in order to have an easier supply of raw material.

The case of Betume shows that artisanal fisheries or fish farming can begin with the establishment of an artisanal fish processing workshop. This initiative, led by women, has received the support of the federal government as well as support from an international cooperation project. Without this support, it would have been very difficult for the women to establish the fish processing workshop and to pay for all the investments needed to begin a viable production process.

THE MARKETING OF SMALL-SCALE FISHERIES PRODUCTS

The three cases presented above show that artisanal fisheries and aquaculture in Latin America are dynamic and diverse. According to its level of organization and to its working means, an artisanal fishing community can sell directly to consumers or to a portfolio of retailers. Artisanal fishermen can also remain isolated and sell their production to the traditional middlemen they know, who control the wholesale marketing at their level. Prices for producers are obviously low when the product passes through two or three successive wholesalers, often poorly preserved with little ice and taking several days before reaching the final consumer. Traceability of fish products is also difficult to track under such conditions.

To a large extent, the marketing of the artisanal fish products in Latin America and elsewhere faces the paradox that quality is inversely proportional to price. The best quality a fish can have is when it comes out of water after being caught by the fisherman. As fish goes through the distribution chain from one wholesaler to another, to retailers and finally to consumers, its quality deteriorates while the price increases. In the end, the consumer receives the worst quality for the highest price while the fisherman sold the best quality fish, for which he received the lowest price.

With few exceptions, the artisanal fisheries production is marketed through the traditional distribution chain, which leads from the production sites to the consumption centres. This distribution chain normally includes two wholesalers, one being responsible for collecting seafood at landing centres and production sites and for transporting it to the seafood wholesale markets in the main consumption centres and the other one being responsible for the sales to retailers at these markets.

Sometimes the second wholesaler is substituted by a processing plant, which produces value-added products. Processing plants must not be confused with wholesalers, who freeze the products they receive and sell them frozen to retailers, who, at their turn, de-freeze the products and sell them as fresh fish products. This second type of processing just adds costs to the product while it lessens its quality. Freezing and de-freezing often do not follow good processing practices. Bad practices of salting and drying of fish products have often similar cost-adding and quality-reducing effects.

Over the last 50 years, Latin American countries experienced a strong migration from rural to urban areas. Today, 80 percent of the 520 million inhabitants of Latin America live in urban areas, most of them in big cities. Over 50 cities in Latin America have more than one million inhabitants. Four cities have more than 10 million inhabitants and two of them, i.e. Mexico City and São Paulo are close to 20 million. The domestic markets in Latin America have become more concentrated from a geographic point of view. INFOPESCA had identified this trend already 10 years ago, when it began to survey the seafood markets in the big Latin American cities. A collection of 14 already published studies gives a broad illustration of Latin American urban markets.

- Mexico City (1998)

  Size of the seafood market: 145 555 tonnes per year; yearly per capita consumption: 8.6 kg; 101 552 tonnes of seafood are consumed fresh.

- Bogotá (2001)

  Size of the seafood market: 42 011 tonnes per year; yearly per capita consumption: 7 kg; 24 237 tonnes of seafood are consumed fresh; supermarkets are responsible for 57 percent of seafood distribution, district markets for 33 percent and restaurants and institutions for 10 percent.

- Caracas (2000)

  Size of the seafood market: 48 478 tonnes per year; yearly per capita consumption: 15.2 kg; 19 200 tonnes of seafood are consumed fresh; fish stores are responsible for 34 percent of seafood distribution, municipal markets for 25 percent, supermarkets for 17 percent, restaurants for one percent and others for 23 percent.

- Maracay (2005)

  Size of the seafood market: 5 870 tonnes per year; yearly per capita consumption: 9.3 kg; supermarkets are responsible for 49 percent of seafood distribution, municipal markets for 29 percent, fish stores for 16 percent, restaurants for 5 percent.

- Valencia (2005)

  Size of the seafood market: 16 836 tonnes per year; yearly per capita consumption: 19.8 kg; supermarkets are responsible for 69 percent of seafood distribution, municipal markets for 15 percent, fish stores for 8 percent and restaurants for 8 percent.

- Recife (2005)

  Size of the seafood market: 26 872 tonnes; yearly per capita consumption: 8.05 kg; supermarkets are responsible for the distribution of 34 percent of all seafood, public markets for 29 percent, restaurants for 6 percent, street markets for 4 percent and others for 27 percent.

- Maceió (2004)

  Size of the seafood market: 12 685 tonnes; yearly per capita consumption: 12.8 kg; municipal markets are responsible for 68 percent of the seafood distribution, supermarkets for 20 percent, street markets for 7 percent and restaurants for 4 percent.

- Aracaju (2004)

  Size of the seafood market: 7 760 tonnes per year; yearly per capita consumption: 16.8 kg; 2 076 tonnes of the seafood are consumed fresh; supermarkets are responsible for 71 percent of seafood distribution, municipal markets for 20 percent, restaurants for 5 percent and fish stores and street markets for 4 percent.

- Brasilia (1997)

  Size of the seafood market: 23 201 tonnes per year; yearly per capita consumption: 12.8 kg; 4 961 tonnes of the seafood are sold fresh; supermarkets are responsible for 59 percent of the seafood distribution, restaurants for 17 percent, institutional catering for 14 percent, street markets for 4 percent and others for 6 percent.

- Rio de Janeiro (1997)

  Size of the seafood market: 167 124 tonnes per year; yearly per capita consumption: 16.4 kg; 54 452 tonnes of the seafood are consumed fresh; supermarkets are responsible for the distribution of 50 percent of all fresh seafood, street markets and ambulant vendors for 25 percent, fish stores for 15 percent, municipal markets for 7 percent and restaurants for 3 percent.

- São Paulo (1998)

  Size of the seafood market: 249 087 tonnes per year; yearly per capita consumption: 15.3 kg; 145 317 tonnes of the seafood are consumed fresh; restaurants and institutions are responsible for 49 percent of all fresh seafood distribution, street markets and municipal markets for 35 percent, "catch & pay" sport fishing establishments for 12 percent and supermarkets for 4 percent.

- Montevideo (1997)

  Size of the seafood market: 12 400 tonnes per year; yearly per capita consumption: 9.1 kg; 5 225 tonnes of seafood are consumed fresh; municipal markets and street markets are responsible for 45 percent of the distribution of fresh fish, fish stores for 32 percent, supermarkets for 12 percent and restaurants for 11 percent.

- Buenos Aires (1997)

  Size of the seafood market: 109 730 tonnes per year; yearly per capita consumption: 9.5 kg; 80 372 tonnes of seafood are consumed fresh; fish stores are responsible for 36 percent of fresh fish distribution, open air markets and municipal markets for 26 percent, supermarkets for 23 percent and restaurants and institutions for 15 percent.

- Santiago de Chile (2000)

  Size of the seafood market: 161 000 tonnes per year; yearly per capita consumption: 26.4 kg; 58 710 tonnes of seafood are consumed fresh; supermarkets are responsible for 45 percent of the seafood distribution, fish stores for 24 percent, restaurants and institutions for 16 percent and open air markets for 15 percent.

This rapid overview of some big Latin American urban markets for fish products shows a great diversity of situations and of annual per capita consumption of fish ranging from 7 kg in the case of Bogotá to 26.4 kg in the case of Santiago. It also shows a diversity of the relative importance of different market segments such as supermarkets, fish stores, restaurants and others.

The challenge of ensuring and preserving the quality of fresh seafood along the distribution chain in Latin America has so far not received the same attention as the challenge of ensuring quality of fish handling, preservation and processing onboard fishing vessels or in fish processing plants. The quality of fresh seafood therefore varies considerably depending on the sales outlet and the place where fish is sold. There are no training programmes implemented or planned for seafood retailers or for wholesalers in Latin America apart from limited efforts made by some international supermarket chains. The relatively low per capita seafood consumption in most cities of the continent is also due to the small number of seafood retail outlets as compared to the number of meat retail outlets for example.

Artisanal seafood products have been exported in various Latin American countries, both as exports of fresh fish as well as of processed fish products. In the latter case, the real clients of the artisanal fisheries sectors are processing plants, which purchase from both industrial and artisanal fishermen as well as from fish farmers. For fresh fish to be exported, first class quality and efficient transport logistics are required. Fresh fish exporters have generally only few customers abroad and supply them with few species. Very often, the exporters are in fact local agents of foreign importers. Prices paid to fishermen for fish destined for export are normally a little higher than for fish destined for the local market in order to encourage fishermen to take special care of their catches, use ice for preservation, handle fish carefully and keep fishing trips as short as possible.

Unfortunately, a common result of the export of fresh fish caught by artisanal fishermen has been in many cases the depletion of the targeted species in fishing grounds used by local fishing communities.

MARKETING POTENTIAL FOR ARTISANAL FISHERIES AND AQUACULTURE PRODUCTS

Most seafood produced by artisanal fishermen and fish farmers passes through two marketing channels. The first channel is the fresh fish market and the second one is the fish processing industry. While there are no reliable statistics available, it is estimated that in Latin America, the fresh fish market is much more important than the processing industry as marketing channel for fish and fish products produced by artisanal fisheries. It is estimated that 70 to 80 percent of the fish produced by the artisanal fisheries and aquaculture sectors passes through the fresh fish market. Most of this fresh seafood is destined for domestic markets but there are some examples, which show that artisanal fish products can also be exported by traders with good transport logistics. Some markets attract seafood produced by artisanal fisheries in neighbouring countries. This is the case in Colombia, where importers often buy fresh seafood from Ecuador, Venezuela or Brazil. Amazonian freshwater fishes caught by artisanal fishermen in Brazil are purchased by wholesalers in Leticia and account for 85 percent of their supplies.

As most coastal fishery resources are already fully exploited, the challenge is not to produce more fish but to sell the same quantities of catches for better prices. The quality of the product supplied by artisanal fishermen is crucial for realizing higher prices. Improving quality requires good logistics for transporting, handling and preserving fish as well as consciousness of the need to maintain the quality of fish along the fish distribution chain. Customers, who are willing to pay for quality, exist everywhere in Latin America. They are normally restaurants, specialized fish shops and some supermarkets. The Women's Association of Betume has understood that better prices can be realized by selling products of higher quality directly to restaurants and that a regular supply is as important as the quality of the product.

The use of ice onboard fishing vessels and at landing sites as well as a quick transport of fish from landing sites to consumption centres are indispensable aspects of improved fish marketing arrangements. There are also other possibilities to realize better prices for artisanal seafood. One of them is to shortcut as many elements as possible in the fish distribution chain and to make use of existing market information. In Mexico, for instance, daily changes of prices at the Nueva Viga wholesale market are easily accessible on the Internet. The same is true for seafood prices at the São Paulo CEAGESP wholesale market. These prices are used as reference prices for any seafood transaction in the country. Access to the Internet is spreading everywhere in Latin America as well as the use of cellular telephones. This helps fishing communities to stay informed and to gain information that can be used in price negotiations. Pricing becomes also more transparent. The fishermen cooperative of Tamiahua, for instance, is aware of the price variations in different markets. Taking advantage of this knowledge, the cooperative sells some of its products locally and sends others to the wholesale markets of Mexico City and Guadalajara. Another way of realizing better prices is to distinguish a product from other, similar products by making the product known and also the community, which produces it, and by highlighting the artisanal character and quality of the product.

As the case studies of Mexico, Peru and Brazil show, there are thousands of artisanal fishing communities in Latin America. These communities are dynamic and each community has its own characteristics. While some fishing communities give up their traditional occupation over time due to urbanization, depletion of traditional fishing grounds, expansion of tourism in coastal regions or other causes, other communities adapt themselves to new developments, grow stronger and get better organized to deal with changes. Most artisanal fishing communities have characteristics, which are already highly valued in present times. These characteristics, when transferred to the products produced by these communities, could enhance the attractiveness of the products for consumers and thereby increase their value. Linking a product to a particular community, which produces it, also conforms to the modern concept of traceability.

Characteristics of artisanal fishing communities, which can be used to differentiate their products from similar products produced by industrial fisheries in a positive way and to highlight product-specific uniqueness, include environmental and cultural characteristics. As the fate of an artisanal fishing community is closely linked to the fate of the living resources in their area, the community is aware of this fact and participates actively in coastal management in order to conserve its aquatic and natural resources, to fight sources of pollution and to ensure the sustainability of its main livelihood activity. This is the case in Tamiahua, where the fishing community uses the resources of a limited and well defined area, i.e. the Tamiahua lagoon. The fishing community and cooperative of Tamiahua carefully monitor the fishery resources and the lagoon ecology and environment and use selective fishing gears. Consumers should appreciate the fact that fish they buy from the Tamiahua Fishermen's Cooperative has been caught in a responsible and sustainable way and originates from a healthy aquatic environment. Artisanal fishing communities can be closest to the concept of responsible fisheries.

Most fishing communities are located in beautiful coastal landscapes, which attract many tourists. Strong communities have taken advantage of tourism, sometimes with the support of their municipalities, as the culture of artisanal fishing communities and the possibility to buy fresh fish directly from the producer are tourist attractions by themselves besides the landscapes, the beaches and the sun. Fishermen do not need to go very far to sell their products and the prices of fish usually follow the higher prices being charged at any tourist site. An example is the case of the fishermen colony of Copacabana beach in Rio de Janeiro, where the municipality has built a small fish store for fishermen to sell their catches. The fishermen colony existed on that beach prior to its urbanization at the beginning of the 20th century. Fishermen repairing their nets, pushing their boats into the water or sorting their catches on return from the sea are part of the landscape of Copacabana. As the case of Copacabana and many fashionable ports from Isla Margarita to Acapulco and from Punta Del Este to Cancún show, tourism and artisanal fisheries can enter into a win-win relationship.

Artisanal fishing communities have many common cultural characteristics such as closeness to sea and nature, seamanship, hard work, solidarity and many more. World literature has always portrayed these characteristics in novels such as "The Old Man and the Sea" by Ernest Hemingway. These characteristics can still be found in many fishing villages in Latin America as well as some characters described by Jorge Amado in his books and in traditional popular songs such as the *jangadeiros* by Bahia's Dorival Caimi.

Further to common characteristics, there are specific cultures related to the national or ethnic origin of artisanal fishermen communities. The national and ethnic cultural heritage of fishing communities can help to differentiate and distinguish their products. For instance, the artisanal fishermen of Mar del Plata in Argentina have their roots in Sicilian immigration and even today, the Sicilian cultural heritage is associated with the artisanal fisheries of this Argentinean harbour. Some of the fish species caught are processed according to the traditional and original techniques such as salted anchovies, which are processed *alici*-style. In Florianópolis, Brazil, artisanal fisheries have been originally developed by Azoreans and some fishing communities still maintain their traditions. Similar traditions and cultural heritages can be found in many communities residing along Latin American coastlines such as Caiçaras in Southern Brazil, Miskitos and Garifunas in Central America, Quechuas and Aymaras on Lake Titicaca and in Mayas on the Yucatan Peninsula. Most of these communities have their unique values, religion, traditions, languages, songs and dances, mythologies, culinary habits and fishing techniques, which can be associated with their products in order to differentiate them from other products and show their uniqueness.

As far as artisanal fish farming is concerned, this is a new activity in Latin America. Most artisanal fish farmers do not live in special communities as artisanal fishermen do. There are exceptions though like the fish farming communities in the lower São Francisco valley. In this case, an effort of differentiation of fish farmed in this area is under way through the adoption of seals of geographical origin. The region is culturally very rich and includes towns having kept their ancient colonial styles. The water of the river used by local fish farmers is particularly clean. Other artisanal fish farmers in Latin American are labelling their products as "organic products" in order to realize higher prices.

CONSTRAINTS TO THE DEVELOPMENT OF SMALL-SCALE FISHERIES

Artisanal fisheries in Latin America face many constraints, which prevent their taking full advantage of the opportunities described above. In many places, artisanal fishermen, particularly part-time and occasional fishermen, are not perceived as professional artisans of fisheries but only as second class fishermen due to the fact that many poor and unskilled persons have taken up fishing because of a lack of other employment opportunities. Artisanal fishermen have typically a low educational level and many of them, particularly the older generation, are still illiterate. Artisanal fishermen are generally considered to belong to the lowest social strata of society.

Many artisanal fishermen do not master their profession well, use inadequate fishing gears, are not aware of the need for resource sustainability and do not care too much about the quality of their products. In Latin American countries, there are few training institutes for artisanal fishermen.

Most of the few existing courses are about the organization and the management of associations or cooperatives, seldom about resource management, selective fishing gears, quality control or seafood marketing. This was not always the case and many older Brazilian artisanal fishermen still remember that during the 1940s and 1950s, there were so called free "fishing boarding-schools" managed by the Navy and especially intended to train artisanal fishermen's sons.

At that time, during and just after World War II, the Navy's aim was to train young men living in fishing communities for the control and defence of the coastline. Nowadays, in peace times in this part of the world, a similar training could be given, for instance, for the control and defence of the marine and riverine environment and ecology.

Domestic seafood distribution and marketing does not receive the same attention from sanitary authorities as exports. The national sanitary services that control the export of seafood are different from those controlling domestic seafood marketing, which often remains under the authority of municipalities. The quality of seafood offered for consumption varies widely. Most countries in Latin America lack training opportunities for seafood vendors. If the quality cannot be assured along the distribution chain, it is unlikely that efforts at only one part of the chain, such as the level of the artisanal fish producer, can yield positive results.

STRATEGIES FOR OVERCOMING CONSTRAINTS TO A FURTHER DEVELOPMENT OF SMALL-SCALE FISHERIES AND THEIR INTEGRATION INTO REGIONAL AND INTERNATIONAL FISH TRADE

The main advantage of Latin American artisanal fisheries is that many artisanal fishing communities are organized in colonies, associations, cooperatives, unions, federations and confederations. It is together with these existing organizations that one can conceive solutions for a better utilization and marketing of the artisanal seafood production.

Many artisanal fishermen organizations are involved in local politics. While some of them might be tempted to increase the number of their members in order to have more political weight, artisanal fishermen organizations are also aware that an increase in the number of fishermen in a community can jeopardize the sustainability of the fishery resources of this community. In some cases, growth of fishing communities can be more gainfully attained through the conversion of part of their activities from capture fisheries to fish farming. An example is the Brazilian state of Santa Catarina, where artisanal fishing communities were encouraged through low interest loans and technical support from state government extension workers to farm mussels and oysters. Over a period of five years, the mussel and oyster production in the state grew from zero to 4 000 tonnes per year. This example could be replicated in many other regions. With the continuous development of aquaculture technology, one can imagine that in a not-too-distant future, many artisanal fishing communities could become involved in sea ranching. This possibility also reinforces the need of organizing communities and involving them in the coastal management and conservation of their region.

It is clear from the above that the main solutions to overcome the existing problems and to take full advantage of existing opportunities for improving the utilization and marketing of artisanal fisheries products are intimately linked to training. A pre-requisite of a sustainable production of high quality seafood, which commands higher prices, is a well organized artisanal fishing community, which masters its profession and is open to innovations. Many modalities of training can be implemented. They require the political will of the concerned governments as well as investments. Technical support and training for fishing communities should focus on better handling of catches, better knowledge of local resources for fish farming and sea ranching and better knowledge of markets and marketing possibilities. Technical support and training can be provided by professional extension workers hired and trained by national or state fisheries authorities. These extension workers can also be backed by local universities working together with the fishermen organizations in their communities. The interaction between fishermen and students of different subjects, i.e. biology, sociology, veterinary medicine, geography, food technology, business administration and other disciplines can only be enriching for both sides.

The establishment of artisanal fishery schools can also play an important role in improving the utilization and marketing of artisanal fish products. Such schools can be established as boarding schools for boys and girls from different fishermen communities, who will be practically taught over a period of a few months about fishing and fish farming techniques, conservation and management of the environment, fish utilization and marketing and other related subjects. Fisheries related training can also be provided at local schools frequented by members of fishing communities in addition to the normal national educational programme. This type of professional training should aim to shape a new, more professional generation of artisanal fishermen and fish farmers with up-to-date knowledge of their profession. Formally supervised apprenticeships are another form of training, which could be considered. In addition to training of artisanal fishermen, training should also be provided for seafood retailers. Training of national trainers in domestic marketing of seafood can be funded by a specialized international organization such as FAO, for instance, and implemented by regional organizations such as INFOPESCA.

Besides training, the improvement of artisanal seafood marketing requires investments in equipment, civil works and working capital. This can only be achieved by organizations like cooperatives, companies or associations, which are legally entitled to market seafood. Investments include investments in ice plants, in workshops for light fish processing such as filleting or peeling, in transport facilities such refrigerated trucks and in working capital. Normally, national and regional investment banks are in a position to provide low interest loans for these types of investment. Often, however, banks do not have the technical background to appraise investment plans prepared and presented by artisanal fishermen cooperatives. In many cases, artisanal fishermen cooperatives do not have the technical background to prepare an investment plan. For this reason, the preparation of investment plans should be assisted by fisheries extension workers. Sometimes, the guarantees required by banks might appear as not affordable to some cooperatives or associations. Affordable interest rates, collateral and guarantee requirements depend on national economic policies, which give priority to these types of investment. In addition to the investment needs of the production side of the distribution chain, one must also foresee investments at the level of wholesalers and retailers in order to assure the supply of quality products to consumers.

As far as legislation and regulations are concerned, most Latin American countries have legislation and regulations that cover artisanal fisheries and seafood distribution. While the definition of "artisanal fisheries" differs from country to country, fisheries legislation including legislation on sanitary matters commonly not only affects artisanal fisheries but also other sectors. The same is true for national legislation regarding the establishment of cooperatives or other categories of associations, legislation regarding the registration of trade marks, seals of indication of geographical origin and other types of legislation.

Normally, artisanal fishermen, their associations and cooperatives and even national fisheries authorities are not aware about all possible implications of the existing national legislation. The preparation and dissemination of national compendiums on the subject among all stakeholders in the artisanal fisheries sector and the entire fisheries and aquaculture sector would be helpful.

The exchange of experiences with legislation and of evaluations of the results of legislation should be organized together with national fisheries authorities in order to assess the adequacy of legislation in relation to national and regional development policies.

While most fishing communities along the Latin American coastline are linked by road to the main consumption centres and are served by electricity, telecommunications including cellular phone coverage and drinkable water, there are still some isolated communities in some parts of the continent which do not benefit from this infrastructure. This is particularly the case in the Amazon, the Andean and the Mosquito Coast regions. With the rapid development of these regions, it is likely that in a couple of decades, basic infrastructure will be available everywhere on the continent.

Bibliography

Acquaah, George: *Fish Marketing*, NJ, Pearson/Prentice Hall, 2004.

Ausubel, F.M.: *Current Protocols in Fish Handling*, New York, John Wiley and Sons, 1989.

Chrispeels, Maarten : *Plants, Genes and Fish Biotechnology*, Sudbury MA, Jones and Barlett Publishers, 2003.

Clark, J.M.: *Submarine Management*, New York, W.H. Freeman and Company, 1977.

Cox, Earl D. : *Beyond Humanity: Cyber Evolution and Future Minds*, Roackland, Charles River Media, 1996.

Curtis, H.: *Introduction of Embryology of Fisheries*, New York, Worth Publishers, 1989.

DeGregori, Thomas R.: *Bountiful Harvest: Technology, Food Safety and the Environment*, Washington DC, Cato Institute, 2003.

Dodds, John H.: *Fish Technology*, New York, Cambridge University Press, 1985.

Fransman M, Junne G, Roobeek A: *The Fish Biotechnology Revolution?*, Oxford, Blackwell, 1995.

Fumento, Michael: *Bioevolution: How Biotechnology is Changing Our World*, San Francisco, Encounter Books, 2003.

Gaskell G: *Biotechnology-the Making of a Global Controversy*, Cambridge, Cambridge University Press, 2002.

Kurzweil, Ray: *The Age of Spiritual Machines*, New York, Penguin Books, 1999.

Moravec, Hans: *Mind Children: The Future of Robot and Human Intelligence*, Cambridge, Harvard University Press, 1988.

Parens, Erik: *Enhancing Human Traits: Ethical and Social Implications*, Washington, D.C.: Georgetown University Press, 1998.

Retzer, W.J.: *Biotechnology and Fishries Technology*, Englewood Cliffs, Prentice Hall, 1991.

Index

□□□